Milestones in Drug Therapy
MDT

Pharmacotherapy of Obesity

Edited by John P.H. Wilding

Birkhäuser
Basel · Boston · Berlin

Editor
John P.H. Wilding
Clinical Sciences Centre
University Hospital Aintree
Longmoor Lane
Liverpool L9 7AL
United Kingdom

Library of Congress Control Number: 2007935857

Bibliographic information published by Die Deutsche Bibliothek
Die Deutsche Bibliothek lists this publication in the Deutsche Nationalbibliografie; detailed bibliographic data is available in the internet at http://dnb.ddb.de

ISBN 978-3-7643-7138-8 Birkhäuser Verlag AG, Basel – Boston – Berlin

Part of Springer Science+Business Media
Printed on acid-free paper produced from chlorine-free pulp. TFC ∞
Cover illustration: see page 82. With friendly permission of Steve R. Bloom.

ISBN 978-3-7643-7138-8 ISBN 978-3-7643-7425-9 (eBook)

9 8 7 6 5 4 3 2 1 www. birkhauser.ch

Contents

List of contributors

Jonathan R. Arch, Clore Laboratory, University of Buckingham, Buckingham MK18 1EG, United Kingdom; e-mail: jon.arch@buckingham.ac.uk

Stephen R. Bloom, Department of Metabolic Medicine, Imperial College London, 6th Floor Commonwealth Building, Hammersmith Hospital, Du Cane Road, London W12 ONN, United Kingdom; e-mail: s.bloom@imperial.ac.uk

George A. Bray, Pennington Biomedical Research Center, LSU System, 6400 Perkins Road, Baton Rouge, LA 70808, USA; e-mail: brayga@pbrc.edu

Owais B. Chaudhri, Department of Metabolic Medicine, Imperial College London, 6th Floor Commonwealth Building, Hammersmith Hospital, Du Cane Road, London W12 ONN, United Kingdom; e-mail: o.chaudri@imperial.ac.uk

John C. Clapham, CVG I Bioscience, Astra Zeneca, Mereside, Alderley Park, Macclesfield SK10 4TG, United Kingdom; e-mail: john.clapham@astrazeneca.com

Joanne A. Harrold, School of Psychology, University of Liverpool, Eleanor Rathbone Building, Bedford Street South, Liverpool L69 7ZA, United Kingdom; e-mail: harrold@liverpool.ac.uk

Muhammad Khan, Diabetes and Endocrinology Clinical Research Group, University Hospital Aintree, Clinical Sciences Centre, 3rd Floor, Lower Lane, Liverpool L9 7AL, United Kingdom

F. Xavier Pi-Sunyer, St. Luke's-Roosevelt Hospital, Columbia University College of Physicians and Surgeons, 1111 Amsterdam Avenue, WH 1020, New York, NY 10025, USA; e-mail: fxp1@columbia.edu

Kirsty L. Smith, Department of Metabolic Medicine, Imperial College London, 6th Floor Commonwealth Building, Hammersmith Hospital, Du Cane Road, London W12 ONN, United Kingdom; e-mail: Kirsty.smith@imperial.ac.uk

John P.H. Wilding, Clinical Sciences Centre, University Hospital Aintree, Longmoor Lane, Liverpool L9 7AL, United Kingdom; e-mail: j.p.h.wilding@liverpool.ac.uk

Preface

In the last 10 years obesity has rapidly moved from being a 'cinderella' branch of general medicine which was largely viewed by health professionals and policy makers as more of a cosmetic than a medical problem, to becoming recognised as an epidemic that now rivals smoking for its adverse effects on health. These include increased risks for many major common diseases including diabetes, cardiovascular disease, respiratory disease, joint disease and many common cancers. Not surprisingly this has led to a race amongst researchers and the pharmaceutical industry to discover new, safe and effective treatments for this common disorder.

The first section of this book sets the scene with a chapter by Xavier Pi-Sunyer on the medical need for obesity drugs in a context of the many medical conditions that can be improved by weight loss. Obesity treatment has a long history and many of the older treatments have been withdrawn or have had their use restricted considerably because of concerns over safety and/or efficacy. In the current challenging regulatory environment it is therefore important to recall the salutary lessons from this early experience of the treatment of obesity which has been expertly reviewed by George Bray. Other than the medical need, the other reason for the explosion in drug development in obesity relates to very rapidly developments that have occurred over the last 10-15 years in our understanding of the regulation of energy balance. This has led to the identification of many new molecular targets with understanding of their role in normal physiology and in the pathophysiology of obesity and related conditions. The prospect the new drugs for obesity can be rationally designed on the basis of sound science is now becoming a reality.

Some of this new science has led to the development of new drugs, three of which have been approved in much of the world in the past 10 years. These include the intestinal lipase inhibitor orlistat, the centrally acting serotonin and noradrenaline reuptake inhibitor sibutramine and the cannabinoid 1 receptor blocker rimonabant.

Preclinical and clinical pharmacology, clinical efficacy and trial data for these drugs is reviewed in section 2, providing the basis for the current pharmacotherapy of obesity.

The final section deals with three broad areas that are the target for much of future drug development. These include drugs acting on the central nervous system, use of peripheral gut hormones and other signals, reviewed by Owais Chaudri, Kirsty Smith and Stephen Bloom and finally peripheral thermogenic targets reviewed by Jonathan Arch and John Clapham.

It is hoped that this book provides the reader with a comprehensive account of the past, current state of the art and likely future developments for pharmacotherapy in obesity.

John P.H. Wilding September 2007

Pharmacotherapy of Obesity
Edited by J.P.H. Wilding

Why drugs?

F. Xavier Pi-Sunyer

St. Luke's-Roosevelt Hospital, Columbia University College of Physicians and Surgeons, 1111 Amsterdam Avenue, WH 1020, New York, NY 10025, USA

Introduction

Drug therapy for obesity has had a difficult past history. A number of drugs have had addictive or toxic properties that have required discontinuation. Pharmacotherapy for obesity has an important role in those persons who have failed behavioral weight loss attempts or as an adjunct to those attempts. The interest in pharmacotherapy for obesity is an outgrowth of the now general recognition that it is a chronic disease that cannot be cured, but can be treated. Treatment, however, will generally be a life-long affair. The focus on drug therapy is due to the frequent failure of non-pharmacological weight loss programs.

At present, only two drugs, sibutramine and orlistat, are approved for long-term use in the US and in much of the rest of the world. The development of new drugs that could help treatment and prevention is greatly needed. The risk/benefit ratio is important in deciding the usefulness of drugs. Drugs are helpful because the defense of baseline body weight by the body is very forceful, no matter what that baseline weight is. Energy expenditure falls and hunger greatly increases when weight is lost. Because of these very strong and sustained defensive biological reactions to weight loss, maintaining weight loss over time becomes increasingly difficult.

There are a large number of possible agents that could be developed. There are a wide variety of neurotransmitters, gut peptides, and other small molecules that are active in food intake and energy expenditure that can be copied or blocked.

It is probable that in the future, as our knowledge base increases, drugs will be developed that will be useful for some persons and not others, according to their individual genomic make-up. That would usher in an era of personalized medicine in the weight loss field.

It is important to accelerate the development of drugs that are safe and effective. Success in this endeavor could prevent a great deal of disease and improve quality of life.

"Diseases desperate grown
by desperate appliance are reliev'd
or not at all" Shakespeare, Hamlet IV.III.9

Historical context

Drug therapy for obesity has been fraught with problems over the years. Early drugs such as amphetamines were found to be addictive and therefore unacceptable [1] (Tab. 1). In the 1950s, phentermine and diethylpropion were developed for weight loss. These drugs, however, were only tested and approved by the FDA for short-term use (less than 3 months) [2] (Tab. 2). Their effect also was modest and they produced significant side effects. In the late 1960s, phentermine was tested for a somewhat longer period (36 weeks) with modest effects [3]. In the 1970s, fenfluramine was introduced, again only approved for short-term use. The weight loss results were, however, somewhat better. In the 1980s, dexfenfluramine, the active component of d,l fenfluramine, was approved and a number of trials demonstrated its efficacy in weight loss [4–7]. It was in this decade that the combination of phentermine and fenfluramine was first tried long-term [8, 9]. Subjects were treated for up to 3.5 years. The obese volunteers were treated with diet, exercise, and behav-

Table 1. History of drug approval by FDA

1950s	1970s	1980s	1990s
Phentermine*	Fenfluramine*	Dexfenfluramine*	Sibutramine†
Diethylpropion*			Orlistat†

* approved for short-term use only
† approved for long-term use

Table 2. Drugs approved for use in the USA

Drug	Drug enforcement administration schedule
Amphetamine†	II
Phenmetrazine†	II
Benzphetamine HCL*	III
Phendimetrazine tartrate*	III
Phentermine HCL*	IV
Diethylpropion HCL*	IV
Mazindol*	IV
(d,l) Fenfluramine‡	IV
Dex fenfluramine‡	IV
Phenylpropanolamine HCL‡	—
Sibutramine	IV
Orlistat	—

† Not recommended for treatment of obesity
‡ Use discontinued
* Approved for short-term use only

ior modification, and were randomized to experimental drugs or placebos. At 60 weeks, patients on continuous treatment had lost 15.8 kg. Individuals who took medication constantly for 3.5 years had persistent weight loss. The efficacy of this combination was thus much greater than had been the case when any of the drugs which were available were used alone, and as a result it was used extensively throughout the world. In the late 1990s reports of toxicity began to surface. These included heart valve abnormalities [10, 11] and primary pulmonary hypertension [12–15]. The fenfluramines were therefore withdrawn from the market.

There was then a lull in the availability of new drugs until the 1990s, when sibutramine and orlistat were introduced. These two drugs are now approved for at least 2 years of use, and in fact physicians are using them for longer periods. Sibutramine is a serotonin and nor-epinephrine re-uptake inhibitor which reduces food intake by enhancing satiety. The drug has been tested in a number of randomized clinical trials and has been found to reduce weight with an average of a 4–8 kg weight loss [16–18]. The other drug is orlistat. Orlistat is an inhibitor of intestinal lipase which impairs fat absorption by the gut. The net effect is to decrease absorption of dietary fat calories. This drug has undergone 2-year clinical trials with no significant side effects except for a small reduction in blood levels of fat soluble vitamins (within the normal range) [19–21]. It is of about the same effectiveness as sibutramine over a 1 year period. In a 1 year placebo-control study, 55% of orlistat-treated patients lost more than 5% and 25% lost more than 10% of their body weight compared to 33% and 15%, respectively, achieving the same mean weight loss in the placebo-treated group. The side effects of this drug are steatorrhea, with soft and more frequent stools. An attempt to use the two drugs in combination did not improve weight loss [22].

There is a school of thought that the modest effect of presently approved drugs and their resultant very low sales are all to the good and that obesity should be treated strictly by diet and exercise and not by drugs. There has been a strange dichotomy in many physicians' and regulators' thinking that, while it is reasonable to have long-term drug therapy for metabolic conditions like high blood pressure, dyslipidemia and diabetes mellitus, it is not acceptable for obesity. This stems from an attitude that obesity is a matter of self-discipline and not a matter of biologic susceptibility. But as more and more is known about the etiology of obesity, it is clear that the enhanced eating behavior and the diminished activity are to a large extent genetically determined [23, 24], and that while environment certainly plays a part, biology is also extremely important.

Why use a drug?

The rationale for the use of a drug for a specific condition includes: (i) the condition predisposes to or exacerbates a disease, (ii) amelioration of the condition improves the disease state or risk, and (iii) the intervention has an acceptable safety profile. Pharmacotherapy for obesity has a role in those who have failed conservative weight loss attempts and is often effective when included in a long-term multi-modal plan.

The interest in pharmacotherapy for obesity is an outgrowth of the now general recognition that it is a chronic disease with genetic underpinnings. The new thinking stresses that a chronic disease cannot be cured but can be treated, and that treatment is a life-long affair and will require medication for life rather than medication for a short period of time. The model for obesity then is diabetes mellitus and hypertension, where chronic medication is an accepted modality of treatment and a cure is not the anticipated result.

The reason that there has been interest in pharmacotherapy for the treatment of obesity is that the attempt to lose weight and particularly to keep it off has been fraught with failure. Data from a number of studies have shown that in behavioral modification programs weight is lost for only the first 4–6 months [25]. After this, weight tends to plateau and then begins to increase. At the end of 4–5 years at a maximum, all of the weight has been regained [26, 27]. There is thus a powerful incentive to find drugs that can improve the success rate in the loss of weight and particularly in the maintenance of this loss.

Risk/benefit ratio

Given the widespread and growing prevalence of obesity, the development of new drugs that could help in treatment and prevention of this condition is greatly needed. What are the characteristics of an ideal anti-obesity drug? It needs to be safe and effective. Safety is pre-eminent since such a drug would be taken by a great number of persons. The risk/benefit ratio is therefore extremely important. What are the risks of using the drug as compared to the risks of not using the drug and either maintaining the elevated weight or actually increasing it? To calculate this for a given drug, one needs to know: (i) how much a given weight gain decreases health and longevity, (ii) whether and how much a given weight loss improves the two, and (iii) how much weight loss warrants treatment with a drug given its side effects. The answer is only known reasonably well on the first of the questions. Obesity reduces health [28], increases mortality [29], and reduces longevity [30, 31]. We know that weight loss reduces many risk factors associated with obesity [25], but we do not know if it actually decreases the incidence of many of the eventual comorbid diseases themselves, although it seems self-evident that it should. So we do not have firm answers to these questions. This is because it is expensive and tedious to do the long-term trials that would be necessary to measure 'events' such as the onset of myocardial infarction, or stroke. We generally only have shorter term studies that have measured risk factors and have shown them to be improved after a period of weight loss [32]. Thus, it has been difficult to judge when an individual drug is worth using. Also, some drugs may just lower weight while others may have additional independent effects on some of the risk factors for diabetes and cardiovascular disease. Some may improve certain risk factors but not others. Some may have more unpleasant side effects than others.

Weight change effects on energy balance

As previously mentioned, it is very difficult to lose weight, and it is even more difficult to keep it off. An individual's body defends the highest weight that has been attained. As soon as one begins to lose weight, two phenomena occur: first, energy expenditure decreases and, second, the urge to eat increases.

Energy expenditure changes

The data on energy expenditure is clear. Early human studies showed that with hypocaloric dieting energy expenditure dropped 10–15% from baseline. Later studies have confirmed this [33]. The mechanisms for this are not totally clear, but the thyroid gland, the sympathetic nervous system, as well as central nervous system changes are involved.

The thyroid makes L-thyroxine (T_4), which is metabolized to L-triiodothyronine (T_3). These two thermogenic hormones help to maintain the normal resting metabolic rate. As calorie intake decreases and weight is lost, the thyroid switches from producing T_3 to producing reverse T_3 (rT_3). Unlike T_3, which is thermogenic, rT_3 is not, so that the stimulus to energy production decreases [34]. Figure 1 shows the thyroid hormone products.

A second hormone that influences energy expenditure is leptin. Leptin is secreted from fat cells and is active in the central nervous system [35–38]. As weight is lost, fat cells decrease in volume and leptin production drops [39].

3, 5, 3', 5' - THYROXINE (T_4)

3, 5, 3'- TRIIODOTHYRONINE (T_3)

3, 3', 5'- TRIIODOTHYRONINE (reverse T_3, rT_3, T_3')

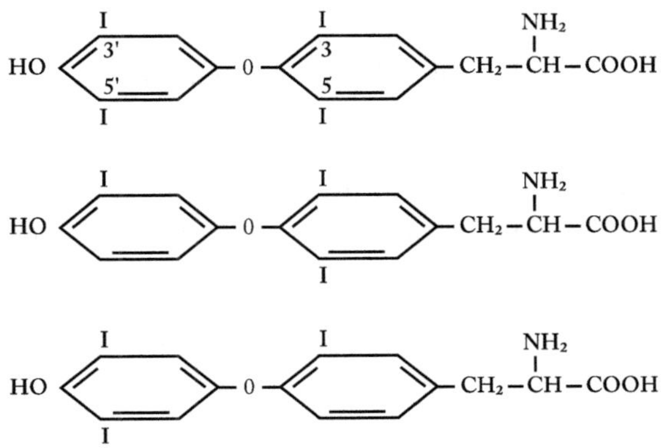

Figure 1. Chemical structures of L-thyroxine (T_4), L-triiodothyronine (T_3), and reverse T_3 (rT_3). From Hershman JM (1980) *Endocrine pathophysiology*. Lea & Febinger, Philadelphia.

Leptin has two primary roles in energy balance. One is to decrease food intake and the other is to increase energy expenditure [37, 39]. With weight loss, leptin levels decrease and this decreases the stimulus to the sympathetic nervous system which, when activated, enhances thermogenesis. The nor-epinephrine released at sympathetic synapses throughout the body is responsible for this. As sympathetic activity falls, energy expenditure drops.

Ingestive behavior

At the same time, the decrease in leptin enhances the urge to eat. There is a school of thought that leptin is not an 'obesity' hormone but an 'under-nutrition' hormone and that its primary action and purpose is to enhance the urge to eat in states of under-nutrition. If so, then the body interprets any drop in weight of an obese person as under-nutrition and defends the baseline weight. Leptin activates neuropeptide Y and agouti-related protein in the central nervous system and stimulates food intake [39].

Because of these two very strong and sustained biological reactions to weight loss, maintaining the loss over time becomes increasingly difficult.

What kind of a drug are we looking for?

Efforts have been directed at finding drugs that will increase energy expenditure, decrease food intake, or prevent absorption of ingested calories. The quintessential drug for treating obesity was thought to be leptin, since this hormone both decreases food intake and increases energy expenditure in rodents which are leptin deficient [37, 38]. When it was first discovered, it was thought to be the answer to the riddle of adequate treatment. However, when it was tested in clinical trials, the result was extremely disappointing. Very little weight was lost and there were significant local side effects [40]. It was found that most obese humans had a resistance to leptin, which has remained unexplained to this day [41]. A close relative of leptin, ciliary neurotrophic factor, was also tested and also found to be wanting [42].

Gut peptides such as cholecystokinin, bombesin, and PYY have been shown to decrease the size of meals [43–45], but have been very difficult to formulate in ways that are easy to administer and also they have not shown a sustained effect. To date, the only gut peptide that seems to have some weight loss potential is exendin-4, an analog of glucagon-like peptide-1 that has been isolated from the saliva of the Gila monster [46].

Other molecules were developed to try to increase energy expenditure. A number of β3 agonists have been tested in humans, but all have had unacceptable cardiovascular side effects related to sympathetic activation. Molecules that activated the uncoupling protein (UCP) system also have been tried. Again, they have been found unacceptable to date.

Are there new potential drugs being developed?

Drugs with new and different mechanisms of action are being developed for obesity. The endocannabinoid system has been discovered and systematically analyzed [47]. This has led to drug development of antagonists to the CB1 receptor, which has been identified as playing a role in food intake regulation [48, 49]. A CB1 endocannabinoid receptor blocker has been developed and has undergone four Phase III trials [50–52]. It is now approved for use in many countries. This drug is promising as one that has somewhat better weight loss to the two drugs now approved for use and also has an impact on elevated blood glucose and lipids. Other CB1 blockers are also being developed.

Other potential drugs under development are looking at various angles for decreasing food intake; for instance agonists of serotonin (5-HT), specifically to the 5-HT2c receptor, as well as agonists to nor-epinephrine, dopamine or histamine. The whole leptin/insulin central nervous system pathway is being analyzed for drug development. Included in this are NPY antagonists, AGRP antagonists, POMC promoters, CART promoters, aMSH analogs, MC4 receptor agonists. Another avenue is to look at gut peptides, such as agents that may increase CCK or PYY activity, GLP-1 activity, decrease ghrelin activity, as well as drugs that could enhance thermogenesis, such as β3 stimulators, UCP homologs, thyroid receptor agonists, UCP3 modulators, and estrone derivatives. Other potential targets are lipase inhibitors with fewer unpleasant gastrointestinal side effects.

The herd *versus* the individual

Given present knowledge, it is simplistic to believe that a specific drug will work to reduce body weight in all obese persons. A particular drug is likely to be effective for some and not for others, depending on the nature of their genetic defect(s). Since drug developers have had little clue of the nature of the defects, it has been difficult to target a specific pharmacological therapy to a specific defect. Since drugs have not been matched to specific etiological conditions, it is certainly possible that a particular drug may be effective for one person but not for another. Going forward, the focus towards a more personalized pharmacotherapy (pharmacogenomics) is likely to sharpen.

Should the search for new drugs be encouraged?

Obesity has become one of the most important public health issues of our time. Using BMI of >25 as overweight and BMI of >30 as obese, over one billion adults worldwide are estimated to be overweight and 300 million or more obese [53]. While the first effort in prevention and treatment of this condition is behavioral lifestyle change to improve nutrition and increase physical activity, many persons do not lose weight, lose very little, or actually increase. Despite much hard work at investigating how to best treat individuals with excess fat the results have generally been disappointing [54]. Only bariatric surgery has been successful, but at the cost of significant side effects [55, 56]. As a result, help in treatment could come through appropriate use of effective and safe drugs that could be used long-term. The development of such agents is badly needed. It is important to accelerate development of drugs that are safe and effective. Success in this endeavor could prevent a great deal of disease and expenditure of money and significantly improve quality of life.

References

1 Griffiths RR, Brady JV, Bradford LD (1979) Predicting the abuse liability of drugs with animal drug self-administration procedures: psychomotor stimulants and hallucinogens. *Adv Behav Pharmacol* 2: 163–208
2 Bray GA (1993) Use and abuse of appetite-suppressant drugs in the treatment of obesity. *Ann Intern Med* 119: 707–713
3 Munro JF, MacCuish AC, Wilson EM, Duncan LJ (1968) Comparison of continuous and intermittent anorectic therapy in obesity. *Br Med J* 1: 352–354
4 Guy-Grand B, Apfelbaum M, Crepaldi G, Gries A, Lefebvre P, Turner P (1989) International trial of long-term dexfenfluramine in obesity. *Lancet* 2: 1142–1145
5 O'Connor HT, Richman RM, Steinbeck KS, Caterson ID (1995) Dexfenfluramine treatment of obesity: a double blind trial with post-trial follow-up. *Int J Obes* 19: 181–189
6 Noble RE (1990) A six months study of the effects of dexfenfluramine on partially successful dieters. *Curr Ther Res* 47: 612–619
7 Finer N, Craddock D, Leveille R, Keen H (1988) Effect of 6 months therapy with dexfenfluramine in obese patients: studies in the United Kingdom. *Clin Neuropharmacol* 11 (Suppl 1): S179–186
8 Weintraub M, Hasday JD, Mushlin AI, Lockwood DH (1984) A double blind clinical trial in weight control: use of fenfluramine and phentermine alone and in combination. *Arch Intern Med* 14: 1143–1148
9 Weintraub M, Sundaresan PR, Madan M, Schuster B, Balder A, Lasagna L, Cox C (1992) Long-term weight control study I (week 0 to 34). *Clin Pharmacol Ther* 51 (suppl): 586–594
10 Connolly HM, Crary JL, McGoon MD, Hensrud D, Edwards BS, Edwards WD, Schaff HV (1997) Valvular heart disease associated with fenfluramine-phentermine *N Engl J Med* 337: 581–588
11 Connolly HM, McGoon MD (1999) Obesity drugs and the heart. *Curr Probl Cardiol* 24: 745–792
12 Brenot F, Herve P, Petitpretz P, Parent F, Duroux P, Simonneau G (1993) Primary pulmonary hypertension and fenfluramine use. *Br Heart J* 70: 537–541
13 Roche N, Labrune S, Braun JM, Huchon GJ (1992) Pulmonary hypertension and dexfenfluramine. *Lancet* 339: 436–437
14 Atanassoff PG, Weiss BM, Schmid ER, Tornic M (1992) Pulmonary hypertension and dexfenfluramine. *Lancet* 339: 436
15 Abenhaim L, Moride Y, Brenot F, Rich S, Benichou J, Kurz X, Higenbottam T, Oakley C, Wouters E, Aubier M et al. (1996) Appetite-suppressant drugs and the risk of primary pulmonary hypertension. *N Engl J Med* 335: 609–616
16 Bray G, Blackburn GL, Ferguson JM, Greenway FL, Jain AK, Mendel CM, Mendels J, Ryan DH, Schwartz SL, Scheinbaum ML et al. (1999) Sibutramine produces dose- related weight loss. *Obes Res* 7: 189–198
17 Bray G, Ryan DH, Gordon D, Heidingsfelder S, Cerise F, Wilson K (1996) A double- blind randomized placebo-controlled trial of sibutramine *Obes Res* 4: 263–271
18 Fanghanel G, Cortinas L, Sanchez-Reyes L, Berber A (2000) A clinical trial of the use of sibutramine for the treatment of patients suffering from essential obesity. *Int J Obes* 24: 144–150
19 Davidson MH, Hauptman J, DiGirolamo M, Froeyt JP, Halsted CH, Heber D, Heimburger DC, Lucas CP, Robbins DC, Chung J et al. (1999) Weight control and risk factor reduction in obese

subjects treated for 2 years with orlistat: a randomized controlled trial. *JAMA* 281: 235–242
20 Sjöström L, Rissanen A, Andersen T, Boldrin M, Golay A, Koppeschaar H, Krempf M (1998) Randomised placebo-controlled trial of orlistat for weight loss and prevention of weight regain in obese patients. *Lancet* 352: 167–172
21 Hauptman J, Lucas C, Boldrin MN, Collins H, Segal HR (2000) Orlistat in the long-term treatment of obesity in primary care settings. *Arch Fam Med* 9: 160–167
22 Wadden TA, Berkowitz RI, Womble LG, Sarwer DB, Arnold ME, Steinberg CM (2000) Effects of sibutramine plus orlistat in obese women following 1 year of treatment by sibutramine alone: a placebo-controlled trial. *Obes Res* 8: 431–437
23 Barsh GS, Farooqi IS, O'Rahilly S (2000) Genetics of body-weight regulation. *Nature* 404: 644–651
24 Barsh GS, Schwartz MW (2002) Genetic approaches to studying energy balance: perception and integration. *Genetics* 3: 589–600
25 Pi-Sunyer FX (1996) A review of long-term studies evaluating the efficacy of weight loss in ameliorating obesity *Clin Therapeut* 18: 1006–1035
26 Wadden TA, Sternberg JA, Letizia KA, Stunkard AJ, Foster GD (1989) Treatment of obesity by very low calorie diet, behavior therapy, and their combination: A five-year perspective. *Int J Obes* 13: 39–46
27 Wadden TA (1993) Treatment of obesity by moderate and severe caloric restriction: results of clinical research trials. *Ann Intern Med* 119: 688–693
28 Pi-Sunyer FX (1983) Medical hazards of obesity. *Ann Int Med* 119: 655–660
29 Flegal KM, Graubard BI, Williamson DF, Gail MH (2005) Excess deaths associated with underweight, overweight, and obesity *JAMA* 293: 1861–1867
30 Fontaine KR, Redden DT, Wang C, Westfall AO, Allison DB (2003) Years of life lost due to obesity. *JAMA* 289: 187–230
31 Peeters A, Barendregt JJ, Willekens F, Nusselder W, Bonneux L, NEDCOM, The Netherlands Epidemiology and Demography Compression of Morbidity Research Group (2003) Obesity in adulthood and its consequences for life expectancy: a life-table analysis. *Ann Intern Med* 138: 24–32
32 Pi-Sunyer FX (1993) Short-term medical benefits and adverse effects of weight loss. *Ann Intern Med* 119: 722–726
33 Leibel RL, Rosenbaum M, Hirsch J (1995) Changes in energy expenditure resulting from altered body weight *N Engl J Med* 332: 621–628
34 Danforth E, Burger AG (1989) The impact of nutrition on thyroid hormone physiology and action. *Annu Rev Nutr* 9: 201–227
35 Zhang Y, Proenca R, Maffei M, Barone M, Leopold L, Friedman JM (1994) Positional cloning of the mouse obese gene and its human homologue. *Nature* 372: 425–432
36 Maffei M, Halaas J, Ravussin E, Pratley RE, Lee GH, Zhang Y, Fei H, Kim S, Lallone R, Ranganathan S et al. (1995) Leptin levels in human and rodent: measurement of plasma leptin and *ob* RNA in obese and weight-reduced subjects. *Nat Med* 1: 1155–1161
37 Pelleymounter MA, Cullen MJ, Baker MB, Hecht R, Winters D, Boone T, Collins F (1995) Effects of the obese gene product on body weight regulation in *ob/ob* mice. *Science* 269: 540–543
38 Campfield LA, Smith FJ, Guisez Y, Devos R, Burnn P (1995) Recombinant mouse OB protein: evidence for a peripheral signal linking adiposity and central neural networks. *Science* 269: 546–549
39 Myers MG Jr (2004) Leptin receptor signaling and the regulation of mammalian physiology. *Rec Progr Horm Res* 59: 287–304
40 Heymsfield SB, Greenberg AS, Fujioka K, Dixon RM, Kushner R, Hunt T, Lubina JA, Patane J, Self B, Hunt P et al. (1999) Recombinant leptin for weight loss in obese and lean adults: a randomized, controlled, dose-escalation trial. *JAMA* 282: 1568–1575
41 Considine RV, Sinha MK, Heiman ML, Kriauciunas A, Stephens TW, Nyce MR, Ohannesian JP, Marco CC, McKee LJ, Bauer TL et al. (1996) Serum immunoreactive-leptin concentrations in normal-weight and obese humans *N Engl J Med* 334: 292–295
42 Ettinger MP, Littlejohn TW, Schwartz SL, Weiss SR, McIlwain HH, Heymsfield SB, Bray GA, Roberts WG, Heyman ER, Stambler N et al. (2003) Recombinant variant of ciliary neurotrophic factor for weight loss in obese adults: a randomized, dose-ranging study. *JAMA* 289: 1763–1764
43 Pi Sunyer FX, Kissileff HR, Thornton J, Smith GP (1982) C terminal octapeptide of cholecystokinin decreases food intake in obese men. *Physiol and Behav* 29: 627–630

44 Muurahainen NE, Kissileff HR, Pi Sunyer FX (1993) Intravenous bombesin reduces food intake in man. *Am J Physiol* 264: R350–R354
45 Batterham R, Cohen M, Ellis S, Le Roux C, Withers D, Frost G, Ghatei M, Bloom S (2003) Inhibition of food intake in obese subjects by peptide YY3–36. *N Engl J Med* 349: 941–948
46 DeFronzo RA, Ratner RE, Han J, Kim D, Fineman M, Baron A (2005) Effects of exenatide (exendin-4) on glycemic control and weight over 30 weeks in metformin-treated patients with type 2 diabetes. *Diabetes Care* 28: 1092–1100
47 Pacher P, Batkai S, Kunos G (2006) The endocannabinoid system as an emerging target of pharmacotherapy. *Pharmacol Rev* 58: 389–462
48 Di Marzo V, Goparaju SK, Wang L, Liu J, Batkai S, Jarai Z, Fezza F, Miura GI, Palmiter RD, Sugiura T et al. (2001) Leptin-regulated endocannabinoids are involved in maintaining food intake. *Nature* 410: 822–825
49 Cota D, Marsicano G, Tschop M, Grubler Y, Flachskamm C, Schubert M, Auer D, Yassouridis A, Thone-Reineke C, Ortmann S et al. (2003) The endogenous cannabinoid system affects energy balance via central orexigenic drive and peripheral lipogenesis. *J Clin Invest* 26: 2442–2450
50 Van Gaal LF, Rissanen AM, Scheen AJ, Ziegler O, Rossner S, Rio-Europe Study Group (2005) Effects of the cannabinoid-1 receptor blocker rimonabant on weight reduction and cardiovascular risk factors in overweight patients: 1-year experience from the RIO- Europe study. *Lancet* 365: 1389–1397
50a Després JP, Golay A, Sjostrom L (2005) Rimonabant in Obesity-Lipids Study Group. Effects on metabolic risk factors in overweight patients with dyslipidemia. *N Engl J Med* 353: 2121–2134
51 Pi-Sunyer FX, Aronne LJ, Heshmati HM, Devin J, Rosenstock J for the Rio-North America Study Group (2006) Effect of rimonabant, a cannabinoid-1 receptor blocker, on weight and cardiometabolic risk factors in overweight or obese patients. RIO-North America: a randomized controlled trial. *JAMA* 295: 761–775
52 Scheen AJ, Finer N, Hollander P, Jensen MD, Van Gaal LF, Rio-Diabetes Study Group (2006) Efficacy and tolerability of rimonabant in overweight or obese patients with type 2 diabetes: a randomized controlled study. *Lancet* 368: 1660–1672
53 World Health Organization. Obesity and overweight facts. Available at: http://www.who.int/mediacentre/factsheets/fs311/en/print.html (accessed 18 January 2006)
54 Yanovski SZ, Yanovski JA (2001) Obesity. *N Engl J Med* 346: 591–602
55 Flum DR, Dellinger EP (2004) Impact of gastric bypass operation on survival: a population-based analysis. *J Amer Coll Surg* 199: 543–551
56 Santry HP, Gillen DL, Lauderdale DS (2005) Trends in bariatric surgical procedures. *JAMA* 294: 1909–1917

Pharmacotherapy of Obesity
Edited by J.P.H. Wilding

Some historical aspects of drug treatment for obesity

George A. Bray

Pennington Biomedical Research Center, Baton Rouge, LA 70808, USA

"A desire to take medicine is, perhaps, the great feature which distinguishes man from other animals"
Sir William Osler [1]

Introduction

Janus was the Roman god with two faces – one face smiling, one scowling [2, 3]. He is frequently seen as the symbol for theater – tragedy and comedy being his two faces. Amphetamine, the first drug with a clinical trial showing effectiveness in the battle of the bulge, also has two faces. One face was suppression of appetite, the other face was addiction. It was this latter frowning face that has cast a cloud of doubt and concern over pharmacological agents developed to help in the battle of the bulge. This is the story.

The report by Lesses and Myerson [4] on the use of amphetamines in the treatment of corpulence was published in the prestigious *New England Journal of Medicine* in 1938. It stands as a landmark in the field of drug development for the battling of the bulge, because it was the first clinical trial of a potential weight loss drug. It provided important lessons on the consequences of use and abuse of drugs for treatment of corpulence and was selected for that reason. Amphetamine and drug addiction are a good example of the 'law of unintended consequences'. Appetite suppression and weight loss were the intended consequences; drug dependency the unintended consequence. I will trace the development and disappointments of using drugs as a strategy for battling the bulge over the past 100 years, since it was just over 100 years ago that drugs were first used in treating the overweight.

Thyroid extract

The introduction of drugs in the battle of the bulge can be dated to at least 1893. The first *bona fide* drug to be used was thyroid extract [5]. In the years before 1890, a clinical condition called myxedema had been clearly identified

[6] to result from failure of the thyroid gland. Patients with myxedema have a puffy type of weight gain, slowing of their thought processes, slow speech, and if severe, a drop in body temperature and coma. When patients with myxedema are treated with thyroid extract, all of these symptoms improve. They became less sluggish, they slept less, and most important for us, they lost weight. Removal of the thryoid gland will produce a myxedema-like picture in animals and humans, and thyroid replacement corrects it, indicating a cause and effect relationship for thyroid deficiency and myxedema. Patients with myxedema are often overweight. The treatment of overweight patients with thyroid extract was empirically supported by the fact that thyroid extract produced weight loss in patients with myxedema. These observations prompted Baron (see Putnam [5]) to use thyroid extract to treat overweight non-myxedematous patients.

Thyroid extract is rich in iodine. These molecules of iodine are attached to thyroxine, triiodothyronine and iodo-tyrosine that are part of a large protein (thyroglobulin) stored in the thyroid gland. When this large molecule is broken down, these iodine containing hormones are released into the blood. Thyroxine was one of the first hormones to be isolated and then synthesized by organic chemists [7, 8]. The thyroxine in thyroid extract can be converted to the active hormone, tri-iodothyronine [9] by chemical changes called deiodination that can occur in several tissues in the body.

Following the invention of a room-sized calorimeter to study human metabolism by Atwater and Rosa in 1895 [10], smaller instruments to measure 'basal metabolism' came into wide clinical use in many hospitals and clinics across America. Thyroid hormone increases basal metabolism and a high or low basal metabolism was widely used in the first half of the 20th Century to diagnose overactivity or underactivity of the thyroid gland that produced hyperthyroidism or hypothyroidism [11]. Measurement of 'basal metabolism' in overweight people was often reported as low [11] although the total amount of oxygen used by overweight patients in this test was higher than normal. The reason tests were often reported as low, when in fact they were not low, is that this test is expressed in relation to body weight. When expressed in relation to body weight, the metabolism of people who were battling the bulge was low. This led to the erroneous concept that obesity might be due to 'low metabolism' and it provided a rationale for treatment of corpulent patients with thyroid hormone. With improved methods measuring metabolic rate that were developed after World War II (1939–1945) we have learned that total energy expenditure in overweight individuals is actually increased [12]; they almost never have low metabolism.

Thyroid extract, and the thyroxine and triiodothyronine that it contains have all been used to treat obesity [5, 13–16]. A major consequence of treatment with thyroid hormone is an increased metabolic rate. When the metabolic rate is increased we burn up, or metabolize fat. We also burn up some of the lean tissue, including muscle and bone that are essential for life [13]. Since obesity is not due to low basal metabolism or low thyroid hormone levels, treatment with thyroid hormone is not indicated unless the patient is also hypothyroid, and suffering from insufficiency of the thyroid gland.

Dinitrophenol

The rapid growth of the chemical industry in Germany in the 19th and early 20th Centuries produced many compounds for dyeing cloth. One major impact of this commercial development was the introduction of dyes to stain histological samples for study of tissue structure. Another outgrowth was a supply of chemicals to the synthetic organic chemist to use for making 'new molecules'. One of the medical pioneers whose work was dependent on these dyes was Paul Ehrlich (1854–1915), the father of drug therapy. Among his many contributions to biomedical science is his concept of the 'magic bullet' – the molecule that would act as the key to a lock and provide a way to targeting chemicals to disease. One of the fruits of his labor was salvarsan also called '707', a drug that saw wide use in the treatment of syphilis before penicillin became available. The number 707 was the number of different molecules he had to test before finding the 'magic bullet' [17].

Another product of this industry that had a direct impact on obesity was dinitrophenol. Factory workers preparing this chemical were noted to lose weight. This finding led doctors to use dinitrophenol in the battle of the bulge without adequate clinical testing. The results were not good – the unhappy side of Janus [18]. Use of dinitrophenol was discontinued after the development of skin rash, cataracts, and neuropathy, but this was only after considerable damage had been done [19].

Amphetamine

After dinitrophenol disappeared as a strategy in the battle of the bulge, in the early 1930s Lesses and Myerson [4] published their paper on the use of amphetamine. Amphetamine was first synthesized by Edeleano in 1887, but it was not until 1927 that Alles described its psychopharmacologic effects. Its two major effects are an increase in alertness, and a decrease in food intake [20]. Trials of this drug as a treatment for narcolepsy, or sleepiness, were initiated in the 1930s and as a pseudo-serendipitous part of these trials it was observed that patients treated with amphetamine lost weight. Following this observation, Lesses and Myerson [4] conducted a clinical trial and demonstrated that amphetamine (Benzedrine®) was effective in producing weight loss. This observation has stimulated controversy and comment ever since.

Amphetamine produces weight loss by reducing food intake [21]. When 10 human subjects were maintained on a constant food intake and treated with amphetamine for 56 days, there was some weight loss in the first 4 to 8 days in seven of the subjects, which was attributed to a slight increase in energy metabolism [21]. Other than that, weight remained stable. When dogs were treated with 5–10 mg of amphetamine just prior to presentation of their daily allotment of food, the drug caused complete abolition of the desire to eat food for a period of 10–21 days in some of these animals. Based on these studies, it was concluded that amphetamine significantly reduced food intake and thus reduced body weight. Following the demonstration that amphetamine suppressed appetite, it was soon realized that the drug was addictive [20]. Appetite suppression and drug abuse are two sides of the same compound – its Janus faces.

After World War II, amphetamine became a street drug that was widely abused and had significant potential for harm. Amphetamine was also widely used in the 1950s by college students to stay awake while studying for examinations and by truck drivers to stay awake when driving long hours. ‘Benzedrine® inhalers’ with amphetamine turned into delivery systems for abusive drugs. In the 1960s restrictive measures curtailed this public health problem, but amphetamine is still an abused drug.

Amphetamine (Benzedrine® chemically identified by the structure alpha-methyl-beta-phenethylamine = amphetamine) is a compound of the β-phenethylamine chemical series. In this way it resembles norepinephrine, an important neurotransmitter in the brain and autonomic nervous system. Amphetamine also resembles ephedrine, another drug that has been used to treat asthma and the battle of the bulge. To learn about the actions of these drugs Barger and Dale, in 1919 [22], evaluated a series of similar drugs to find out what they did and what parts of the molecule were important for this effect. They found that a number of these chemical compounds could mimic the effects on the body of the sympathetic nervous system; hence they are called ‘sympathomimetic’ amines. This work was largely forgotten until more than a decade later, when ephedrine was ‘rediscovered’ to produce sympathomimetic effects, including dilatation or enlargement of the pupils, constriction of the bronchi in the lungs with wheezing, constriction of blood vessels with a rise in blood pressure, and stimulation of the heart rate [20].

Because amphetamine stimulates wakefulness, several studies in the mid-1930s examined the effect on narcolepsy, a state of increased sleepiness. In a study of nine cases of narcolepsy treated with amphetamine, Prinzmetal and Bloomberg [23] did not report any effects on body weight. Similarly, Myerson [24] in his first report on the treatment of fature using Benzedrine® in normal and neurotic persons did not note any weight loss. However in 1937, Nathanson [25] reported a study of 40 patients and noted that ten of them had a marked loss of appetite with a definite reduction in body weight. Losses of weight varied between 3.2 to 9 kg (7 and 20 lbs). The loss of weight appeared to be explained by the lessened appetite and increased physical activity. Davidoff and Reifenstein [26] concluded from their studies with amphetamine that “it may be of use in reducing weight” – an aid in the battle of the bulge. Ulrich [27] in his report on the treatment of narcolepsy with Benzedrine sulfate also noted that seven obese patients lost weight. It is against this background that the clinical study of Lesses and Myerson [4] was conducted. They studied 17 overweight patients from their private medical practice. Each individual was given a 1,400 kcal/d diet and provided with enough medication to last until their next visit two weeks later. During the observation period which lasted from 6–25 weeks, patients lost an average of 0.66 kg/wk (1.45 lb/wk). A promising beginning, but for the Janus properties of amphetamine.

It was not long before the abuse potential was recognized and the drug's use for obesity came under a cloud of disapproval. This Janus side of the problem stimulated pharmaceutical chemists to synthesize other drugs that would reduce hunger but not have the potential that amphetamine had. A variety of new drugs were made, tested and marketed, but with the drug abuse epidemic of the 1960s, all of the derivatives of amphetamine have been tarred with the same brush. As amphetamine fell from grace, a similar pall fell over the entire class of compounds for better or for worse and whether deserved or not. As we will see, one of these 'derivatives' had no abuse potential at all, yet was regulated by the US Government as though it did.

To their credit, the pharmaceutical chemists developed compounds that reduced or eliminated the risk of habituation, yet retained appetite-suppressing properties. One of these provides a particularly important lesson in the semantic pitfalls of tarring all compounds that look alike structurally with the same mechanism of action. It is now known, through the work of Leibowitz and Rossakis [28], that direct hypothalamic injections of amphetamine will significantly reduce food intake. This effect involves the release of two neurotransmitters, norepinephrine and dopamine, that act on adjacent cells to activate signals that tell animals they are not hungry. In all likelihood, it is the response to dopamine that is associated with the risk of habituation. Three different groups of chemical compounds were developed as the result of this work by the organic chemists.

Norepinephrine reuptake inhibitors

The first group of compounds was similar to amphetamine, but had lower or very much lower abuse potential, yet retained the appetite-suppressing effects so needed in the battle of the bulge. For this group of compounds, the mechanism of action appeared to be the release of the neurotransmitter, norepinephrine, but not dopamine, from nerve endings in the brain. More recent data suggests that the higher extraneuronal level of norepinephrine may be due to blockade of re-entry into the neuron rather than simply metabolized.

A second group of compounds in this group, typified by mazindol, arose from the observation that a tricyclic inhibitor of norepinephrine reuptake could reduce food intake. Tricyclic drugs provide an important group of drugs for treatment of depression. Indeed the weight loss produced by mazindol was detected during its tests as a potential antidepressant drug. This relationship between drugs that affect depressed mood and weight loss has been observed several times. Two widely used antidepressants, fluoxetine (Prozac®) and sertraline (Zoloft®) produce weight loss. Venlafaxine (Effexor) is an antidepressant that is very similar chemically to a weight loss drug, sibutramine, which was also identified during trials for an antidepressant.

Serotonergic agents

The third compound has structural similarities to amphetamine, but acts by a totally different mechanism. This molecule, d,l-fenfluramine, works by releasing serotonin and partially blocking its reuptake into nerve endings. On paper, d,l-fenfluramine is similar to amphetamine, but it differs in a major way from other derivatives of amphetamine. Whereas treatment of animals with most of the derivatives of amphetamine reduces brain norepinephrine, treatment with fenfluramine does not. Rather, d,l-fenfluramine reduces brain serotonin, whereas other amphetamine derivatives do not. Its mechanism of action is through serotonergic receptors, not norepinephrine receptors. Thus, it is very different in the way it works and is without addictive potential.

This discovery opened a whole new area of research into serotonergic agents as drugs for use in the battle of the bulge. Serotonin is also involved in depression. This was a problem for some patients taking d,l-fenfluramine. When the drug was stopped abruptly, many patients experienced a mild depression. To avoid this, the drug is usually discontinued after several days.

Most of the chemical modification of the amphetamine molecule mimic the effects of activation of the sympathetic nervous system but have little or no risk of habituation or abuse since they do not affect the dopaminergic neurotransmitters that are so often associated with drugs of abuse.

Fenfluramine/phentermine

Serendipity is an important ingredient in human progress. Serendipity refers to making discoveries that aren't directly related to your major efforts. There are two kinds of serendipity [29]. The first is true serendipity, in which the discovery has no relationship to the usual activity of the individual. Three such examples would be the discovery of the Rosetta Stone by Napoleon's army engineers in Egypt, the discovery of the Dead Sea Scrolls by boys playing in caves in Israel, and the discovery of the Lascaux Caves in France by young boys playing in the mountainous areas of Southern France. In each case, the individual making the discovery had not been trained in scientific disciplines or for 'discovery'.

Pseudo-serendipity is a second sort of serendipity. It occurs to individuals who are highly trained in their field, but who make accidental but often momentous discoveries. The discovery of the explosive TNT by Alfred Nobel and the discovery of x-rays by Wilhelm Roentgen are two good examples of accidental discoveries by trained minds, but in areas that were unrelated to their primary search. The observation by Lesses and Myerson that opened up the field of appetite suppressants in the treatment of obesity extended the pseudo-serendipitous observation that a drug being tested for narcolepsy could reduce appetite and body weight.

The reports of cardiac disease (valvular insufficiency) [30] associated with the use of the fenfluramine drugs is an example of the 'law of unintended consequences' which has plagued the treatment of obesity. This disaster also adds to the concern and doubt that surrounds the use of drugs as aids in the battle of the bulge. The story began in the 1980s with a scientific hypothesis. By that time it was well known that fenfluramine acted on the serotonin system to reduce food intake and that the other appetite suppressants acted on the norepinephrine (noradrenergic) system. It was a logical question to ask whether combining drugs acting on each of these two different receptor systems would produce greater weight loss or have fewer side effects than a single drug alone. This was initially tested in a pilot study and then in a drug trial sponsored for 4 years by the National Institutes of Health [31]. In this study the outcome supported the initial idea. Corpulent participants treated with both drugs lost more weight than in trials with single agents alone, and in many cases were able to keep this weight off for more than 3 years during this study of the battle of the bulge.

When the dramatic weight loss results of combining fenfluramine and phentermine became known, the use of 'Fen/Phen', as it was widely called, exploded across the country. Patients and doctors alike were thrilled with the results of using Fen/Phen. For the first time it appeared that corpulent Americans were winning the battle of the bulge as they had rarely done before. Offices dispensing Fen/Phen opened up all over the country. Then came the surprise – or rather the calamity. In July 1997 the first cases of valvular heart disease in patients taking Fen/Phen were reported [30]. Urgent meetings by the US Food and Drug Administration (FDA) assembled enough information to convince them that up to 30% of the patients treated with Fen/Phen might develop valvular heart disease. On 15 September 1997 fenfluramine and dexfenfluramine were pulled from the market worldwide. The Fen/Phen success had been shattered by the 'law of unintended consequences' and added another sad ending to a therapy that offered such promise in the battle of the bulge.

This is not the first disaster to befall overweight patients treated for their obesity. Table 1 provides several more examples. We can see that even with thyroid hormone, first used more than 100 years ago, there were unwanted and potentially hazardous problems. This litany of difficulties associated with many treatments should give us pause when new ones are developed. Careful testing is most important to keep the smiling Janus face looking down on us.

The litany of undesirable effects of anti-obesity treatments has continued to the present. The α-adrenergic agonist, phenylpropanolamine, was removed from the market by the US FDA because there was a small increased risk of stroke in younger women. This relationship did not appear to be dose-dependent and did not occur in men.

Table 1. Unintended or undesired consequences from the use of medications to treat obesity

Year	Drug	Consequence
1892	Thyroid extract	Hyperthyroidism
1932	Dinitrophenol	Cataracts; Neuropathy
1937	Amphetamine	Drug Addiction
1968	Rainbow pills	Deaths due to arrhythmias from pills with thyroid, digitalis and diuretics
1985	Very low calorie diets	Deaths due to arrhythmias related to use of gelatin as major protein source
1997	Fenfluramine and dexfenfluramine	Aortic regurgitation
1998	Phenylpropanolamine	Stroke
2003	Ma huang	Heart attacks and stroke from the ephedra alkaloids

Summary

This chapter focuses on one aspect of the history of drug treatment for obesity. The first drug in this list was thyroid extract that was originally tried more than 100 years ago and was used for more than 70 years. Dinitrophenol, developed by the aniline dye industry, produced weight loss but was associated with cataracts and neuropathy. Amphetamines were the third group of drugs, but had the draw-back of producing addiction. The introduction of derivatives actually worked on the serotonin system. Fenfluramine and its dextro isomer, dexfenfluramine, were used to threat obesity until it was recognized that they produced valvular heart disease leading to their withdrawal from market. This problem of 'unintended' side effects of detrimental effects has been a recurring theme in the history of drug treatment for obesity.

References

1 Osler W (1891) Recent Advances in Medicine. *Science* 17: 170–171
2 Temkin O (1977) The double face of Janus. In: *The double face of Janus and other essays in the history of medicine.* Johns Hopkins University Press, Baltimore, 3–37
3 Koestler A (1978) *Janus. A summing up*. Random House, New York
4 Lesses MF, Myerson A (1938) Human autonomic pharmacology XVI: benzedrine sulfate as an aid in the treatment of obesity. *N Engl J Med* 218: 119–124
5 Putnam JJ (1893) Cases of myxoedema and acromegalia treated with benefit by sheep's thyroids: recent observations respecting the pathology of the cachexias following disease of the thyroid; clinical relationships of Grave's disease and acromegalia. *Am J Med Sci* 106: 125–148
6 Gull WW (1873–1874) On a cretinoid state supervening in adult life in women. *Trans Clin Soc Lond* 7: 180–185
7 Kendall EC (1915) The isolation in crystalline form of the compound containing iodine, which occurs in the thyroid; its chemical nature and physiologic activity. *JAMA* 64: 2042–2043

8 Harington CR (1926) Chemistry of thyroxine I. *Biochem J* 20: 293–313
9 Gross J, Pitt-Rivers RV (1953) 3:5:3'-Triiodothyronine. I. Isolation from thyroid gland and synthesis. *Biochem J* 53: 645–650
10 Atwater WO, Rosa EB (1899) Description of a new respiration calorimeter and experiments on the conservation of energy in the human body. Bulletin of the US Department of Agriculture Office of Experimental Stations, #63
11 DuBois EF (1924) *Basal metabolism in health and disease.* Lea and Febiger, Philadelphia
12 Ravussin E, Lillioja S, Anderson TE, Christin L, Bogardus C (1986) Determinants of 24-hour energy expenditure in man: methods and results using a respiratory chamber. *J Clin Invest* 78: 1568–1578
13 Bray GA, Melvin KEW, Chopra IJ (1973) Effect of triiodothyronine on some metabolic responses of obese patients. *Am J Clin Nutr* 26: 715–721
14 Edwards DAW, Sawyer GIM (1950) The comparative values of dextroamphetamine sulphate, dried thyroid gland and a placebo in the treatment of obesity. *Clin Sci* 9(2): 115–126
15 Gelvin EP, McGavack TH (1949) Dexedrine and weight reduction. *NY State J Med* 49: 279–282
16 Lyon DM, Dunlop DM (1932) The treatment of obesity: a comparison of the effects of diet and of thyroid extract. *Quart J Med* 1: 331–352
17 Ehrlich P, Hata S (1910) *Die experimentelle Chemotherapie der Spirillosen.* Julius Springer, Berlin
18 Bray GA (1976) *The obese patient. major problems in internal medicine.* W.B. Saunders Company, Philadelphia
19 Simkins S (1937) Dinitrophenol and desiccated thyroid in the treatment of obesity: a comprehensive clinical and laboratory study. *JAMA* 108: 2110–2193
20 Leake CD (1958) *The amphetamines. Their actions and uses.* Charles C. Thomas, Springfield, IL
21 Harris SC, Ivy AC, Searle LM (1947) The mechanism of amphetamine-induced loss of weight. A correlation of the theory of hunger and appetite. *JAMA* 134: 1468–1474
22 Barger G, Dale HH (1910) Chemical structure and sympathomimetic action of amines. *J Physiol* 41: 19–59
23 Prinzmetal M, Bloomberg W (1935) The use of benzedrine for the treatment of narcolepsy. *JAMA* 105: 2051–2053
24 Myerson A (1936) Effect of benzedrine sulfate on mood and fatigue in normal and in neurotic persons. *Arch Neurol Psych* 36: 816–822
25 Nathanson MH (1937) The central action of beta-aminopropylbenzene (Benzedrine). *JAMA* 108: 528–531
26 Davidoff E, Reifenstein EC Jr (1937) The stimulating action of benzedrine sulfate: a comparative study of the responses of normal persons and of depressed patients. *JAMA* 108: 1770–1776
27 Ulrich H (1937) Narcolepsy and its treatment with benzedrine sulfate. *N Engl J Med* 217(18): 696–701
28 Leibowitz SF, Rossakis C (1978) Analysis of feeding suppression produced by perifornical hypothalamic injection of catecholamines, amphetamines and mazindol. *Eur Pharm* 53: 69–81
29 Roberts RM (1989) *Serendipity. Accidental discoveries in science.* John Wiley and Sons, New York
30 Connolly HM, Crary JL, McGoon MD, Hensrud DD, Edwards BS, Edwards WD, Schaff HV (1997) Valvular heart disease associated with fenfluramine-phentermine [see comments] *N Engl J Med* 337: 581–588
31 Weintraub M (1992) Long-term weight control, the National Heart, Lung and Blood Institute funded multimodal intervention study. *Clin Pharmacol Ther* 51(5): 581–585

Pharmacotherapy of Obesity
Edited by J.P.H. Wilding

Regulation of energy balance – towards rational drug design in obesity

Joanne A. Harrold[1] and John P.H. Wilding[2]

School of Psychology[1] and School of Clinical Sciences[2], University of Liverpool, Liverpool, L69 7ZA, UK

CNS regulation of energy balance and metabolism

Body weight and fat mass are tightly regulated around a 'set point'. This control of energy balance depends critically on the central nervous system (CNS). The brain regulates many aspects of energy homeostasis, adjusting both the drive to eat and energy expenditure in response to the status of body energy stores and the availability of food. The CNS regions that control energy homeostasis are accessible to numerous circulating hormones and other factors. Within the CNS itself are specific neuronal populations that recognise these signals and act in a network to integrate the multiple inputs, and help determine energy intake and expenditure.

Much progress has been made in identifying the nutritional signals and their neuronal targets and characterising the mechanisms by which these signals are perceived and integrated by the CNS. Through the identification of novel therapeutic targets, the information obtained is now leading to the design of drugs to treat nutritional disorders, particularly obesity, in humans. This chapter reviews some of the major nutritional signals that regulate energy balance and the key neuronal pathways on which they impact. Throughout, the aim is to summarise current knowledge and to discuss the potential of exploiting these targets in the treatment of obesity.

Neuroanatomy of energy homeostasis

Several brain regions are involved in controlling energy homeostasis in mammals. The primary location is the hypothalamus (Fig. 1); others include the amygdala and nucleus accumbens of the forebrain and also structures in the floor of the 4th ventricle.

The arcuate nucleus (ARC) is located in the floor of the 3rd ventricle where it lies immediately above the median eminence (ME). It contains functionally discrete populations of neurones. In one population of ARC neurones the orexigenic neuropeptides neuropeptide Y (NPY) and agouti-gene related peptide

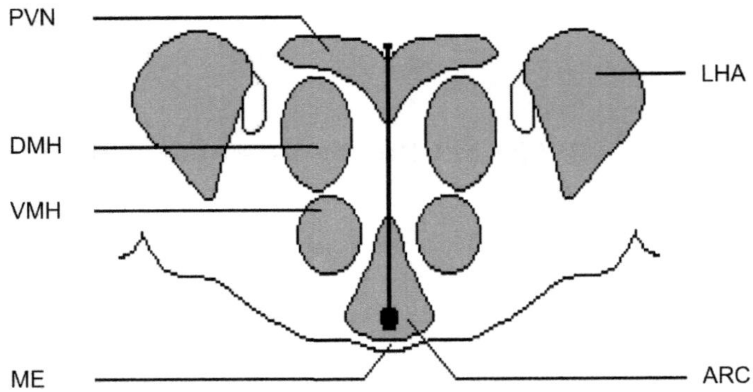

Figure 1. Coronal section of the rat hypothalamus showing the relative positions of the key appetite regulating nuclei with respect to each other through the brain. Key: DMH, dorsomedial nucleus; PVN, paraventricular nucleus; VMH, ventromedial nucleus; ARC, arcuate nucleus; LHA, lateral hypothalamic area; ME, median eminence.

(AGRP) are co-localised. The anorexigenic neuropepetides pro-opiomelanocortin (POMC) and cocaine and amphetamine related transcript (CART) are co-localised in an adjacent subset of ARC neurones [1, 2]. Due to a lack of tight junctions in the capillaries of the ARC and ME, these areas effectively lie outside the blood brain barrier (BBB). Consequently, the ARC neurones are readily accessible to circulating signals of nutritional status. This explains the abundance of receptors for leptin, insulin and the glucocorticoids in the mediobasal hypothalamus. The ARC also has extensive reciprocal connections with other hypothalamic appetite regulating regions including the paraventricular (PVN), ventromedial (VMH) and dorsomedial (DMH) nuclei and the lateral hypothalamic area (LHA).

The PVN is located at the top of the 3rd ventricle in the anterior hypothalamus. It functions to integrate signals from many neuronal pathways important in the regulation of energy homeostasis. These include NPY/AGRP and POMC/CART projections from the ARC and projections from orexin neurones of the LHA [3, 4]. Within the PVN are synaptic terminals which also contain a variety of appetite-regulating neurotransmitters including NPY, α-melanocyte stimulating hormone (α-MSH), serotonin (5-HT), noradrenaline and the opioid peptides. Neuronal pathways also project from the PVN to the vagal nuclei of the medulla, which in turn innervates the Islets of Langerhans [5]. Injection of various neurotransmitters into the PVN influences insulin and glucagon secretion, while lesions of this region result in hyperphagia, reduced energy expenditure and obesity [6].

The VMH is one of the largest nuclei in the hypothalamus. Following classical experiments demonstrating that stimulation of the nucleus inhibited feeding and lesions of the region resulted in hyperphagia and weight gain, the VMH was long considered to be a 'satiety centre' [7]. However, this hypothesis is now known to be oversimplistic as feeding behaviour undoubtedly involves integration of signals arising in several key nuclei. Recent studies have indicated a high abundance of the long isoform of the leptin receptor (Ob-Rb) in neurones of the VMH suggesting that this region is an important target for circulating leptin [8]. The VMH has direct connections with the PVN and DMH; via these connections it may also connect indirectly with the LHA.

The LHA was viewed classically as a 'feeding centre', whose actions opposed those of the VMH. This nucleus comprises a large diffuse population of neurones including subpopulations that express the orexigenic peptides orexin and melanin-concentrating hormone (MCH). NPY neurons form abundant synapses onto the orexin and MCH cell bodies in the LHA [9, 10]. The LHA is also rich in NPY Y5 receptors proposed to mediate the appetite-stimulating effects of NPY [11]. This nucleus is also rich in glucose-sensitive neurones that also respond to insulin [12].

The DMH is located immediately dorsal to the VMH. ARC NPY/AGRP neurones terminate in the DMH [13]. The DMH has extensive connections with other hypothalamic nuclei, such as the PVN. It is thought that the DMH and PVN act as a functional unit to integrate opposing signals from the VMH and LHA. The DMH is also abundant in both insulin and leptin receptors.

The regulation of energy homeostasis also involves multiple brain regions outside the hypothalamus. The floor of the 4th ventricle contains important structures. The nucleus tractus solitarius (NTS) in the medulla plays a significant role. Vagally transmitted gustatory signals (including taste, gastric distension and portal-vein glucose levels) converge on the NTS. The intestinal peptide cholecystokinin (CCK), which is involved in meal termination, also signals to the NTS via receptors on the vagal nerve. Leptin receptors are expressed in the NTS as this area is accessible to circulating signalling molecules, such as leptin, Some NTS neurones also express POMC and melanocortin-4 receptors (MC4-R) [14] and administration of MC4-R agonists and antagonists into the 4th ventricle (adjacent to the NTS) or the lateral ventricle of the hypothalamus elicit similar feeding responses [15]. It has been observed that destruction of the NTS leads to elevated consumption of palatable food [16]. This indicates that the NTS and hypothalamus both play important roles in the processing of information that controls energy homeostasis.

Necessarily, much of the work focusing on the regulation of energy balance derives from rodents. The following sections largely deal with the rodent hypothalamus. Nevertheless, the organisation of the human hypothalamus is broadly similar to that of the rat and much of the information may be directly applicable to the understanding of human appetite and body-weight regulation.

Hypothalamic pathways and neurotransmitters

Within the CNS and particularly the hypothalamic appetite regulating neurones, a variety of products (currently over 50) function to regulate energy homeostasis. These include classical neurotransmitters such as the monoamines (noradrenaline, adrenaline, dopamine and 5HT) and acetylcholine, together with an increasing list of peptides. More recently, some less obvious mediators of appetite such as the lipid based endocannabinoids have been identified. Some of these factors thought important in controlling feeding and metabolism are highlighted in Table 1 and will be discussed in more detail below.

Table 1. Selected appetite-modifying peptides and neurotransmitters, illustrating their central effects on energy balance

Peptide	Effects on energy balance		
	Feeding	Thermogenesis	Body weight
NPY	↑↑	↓	↑
MCH	↑	?	↑*
Orexin A	↑	→	→
Opioids	↑	↓	↑
α-MSH	↓	↑	↓
5-HT	↓	↑	↓
Dopamine	↑	↓	↑
Endocannabinoids	↑↑	↓+	↑

* Effects observed only when animals fed high-fat diet
+ Hypothermic response does not involve brown adipose tissue

CNS hunger peptides

Neuropeptide Y (NPY)

NPY is a 36 amino acid neurotransmitter. It is one of the most abundantly and widely distributed neurotransmitters in the mammalian brain [17]. Anatomical mapping has demonstrated high concentrations of NPY in the hypothalamic appetite-regulating nuclei and in particular within the ARC, where most hypothalamic NPY is derived from.

The administration of NPY to rodents has many experimental effects, but its effects on energy homeostasis are particularly robust. NPY injected into the lateral ventricles or discrete hypothalamic nuclei (PVN or LHA) induces profound hyperphagia (increasing the size and duration of the first meal rather than meal number). These effects of NPY combined with its insulin secretagogue actions and reductions in thermogenesis lead to enhanced energy storage and ultimately obesity [18].

The hyperphagic effects of NPY are mediated by specific NPY receptor subtypes. To date, six receptor subtypes (Y_1–Y_6) have been cloned and characterised. Present understanding suggests that Y_1 receptors mediate the effects of NPY on meal size, although much attention has recently been placed on the Y_5 receptor, which is assumed to serve the role of the NPY 'feeding receptor'.

However, the initial Y_5 antagonists on which this was based have now been shown to be non-selective and to lack major effects on feeding and body weight in rodents [19]. Thus, the involvement of each NPY receptor subtype in the regulation of feeding remains open for debate.

ARC NPY neurones show inappropriate and unrepressed overactivity in animal models of genetic obesity that are caused by leptin receptor mutations (*db/db* mouse and *fa/fa* rat) or the loss of leptin (*ob/ob* mouse). Overactivity of these neurones plays a role in the hyperphagia and reduced energy expenditure that leads to obesity in these models. NPY neurones are also overactive in animals that have lost body weight as a consequence of energy deficit, such as starvation, lactation or insulin-deficient diabetes [20]. The NPY neurones may be stimulated in these conditions by falls in insulin and/or leptin levels, both of which inhibit NPY gene expression [21]. By contrast, NPY neurones are inhibited in dietary-obese animals [22], possibly in response to increased plasma levels of leptin and/or insulin. This suggests an attempt to limit hyperphagia and weight gain.

As yet, it is not clear how important NPY neurones might be in the regulation of human energy homeostasis. If this pathway is already inhibited in obese individuals (as in dietary-obese rodents) then even highly potent and selective NPY antagonists might be ineffective in the treatment of human obesity.

Melanin-Concentrating Hormone (MCH)

MCH is a 19-amino acid cyclic neuropeptide originally isolated from salmon pituitaries. A high-degree of homology is shared between rat, mouse and human MCH and the original salmon peptide [23–25]. In rodents and humans, the prepro-MCH precursor molecule is expressed in a discrete population of neurones located in the zona incerta and LHA [26, 27]. MCH lies within dense core vesicles and in rodents and humans is often co-expressed with CART [28, 29]. MCH containing neurones project widely throughout the CNS suggesting an involvement in numerous physiological functions, including the regulation of energy balance.

Two MCH receptors have been identified. MCH was originally associated with an orphan G-protein-coupled receptor (GPRC) termed SLC-1, although this is now referred to as MCHR1. The structure of MCHR1 is highly conserved across rodents and higher mammals [30] and receptor mRNA and protein levels are widely distributed throughout the brain in a pattern consistent with that of the terminal fields of MCH neurones. High-levels have been found in the nucleus accumbens, amygdala, hippocampus and various hypothalamic appetite-regulating nuclei [31, 32]. With this distribution MCHR1 is likely to mediate the orexigenic effects of MCH in conjunction with other feeding-related functions such as taste, reward and olfaction. The second MCH receptor, termed MCHR2, has only a 38% homology with MCHR1. Furthermore, functional MCHR2 has not been identified in rodents, but is present in human, monkey, ferret and dog [30, 33]. In these species its expression is lower and more restricted than that of MCHR1 [34]. Given these expression characteristics the role of MCHR2 in the regulation of feeding and energy balance remains unclear.

The most widely investigated role for MCH is in the regulation of energy homeostasis. This role was first suggested by the observation of elevated MCH mRNA and peptide levels in the hypothalamus of *ob/ob* mice and *fa/fa* rats as well as fasted animals [35]. Additionally, chronic central infusion of MCH to rodents results in persistent hyperphagia and enhanced body weight and adiposity [35, 36]. Repeated central injections to satiated rats also produce a rapid and dose-dependent increase in food intake. Intriguingly however, these effects are transient and tolerance develops after 5 days of treatment. Chronic overexpression of MCH leads to hyperphagia and an enhanced susceptibility to dietary obesity [37]. By contrast, MCH null mice are hypophagic, with decreased fat mass and increased energy expenditure [38]. Interestingly, mice lacking MCH and leptin are leaner compared to *ob/ob* mice despite eating comparable amounts of food. This is accounted for by enhanced energy expenditure and glucose tolerance [39]. A similar hypermetabolic phenotype is observed in MCHR1 null mice; despite being hyperphagic they are lean and have decreased leptin and insulin levels [40, 41]. The hyperphagia observed in these animals may represent a compensatory response to weight loss.

Given the evidence for a role for MCH in the regulation of feeding, and the characteristics and expression pattern of MCHR1, the latter has been recognised as an attractive target for the development of anti-obesity drugs. Two selective MCHR1 antagonists (T226296 and SNAP-7941) have both been found to suppress feeding induced by central administration of MCH [42, 43]. Furthermore, SNAP-7941 reduces consumption of palatable food and with chronic administration lowers body weight in dietary obese animals. However, the wide distribution of MCHR1 in the CNS indicates the likelihood of multiple roles for MCHR1 signalling. The ability to separate feeding-related effects from other actions will play a large part in determining whether MCHR1 antagonists will represent effective anti-obesity agents in the future.

CNS satiety peptides

Melanocortins and Agouti-Gene Related Peptide (AGRP)

The melanocortin system has come under intense pharmacological scrutiny for its various potential physiological roles including sexual function, pigmentation and energy homeostasis. The melanocortin neurones produce numerous peptides all of which are derived from a common precursor, POMC. Of these, α-melanocyte stimulating hormone (MSH) and β-MSH inhibit feeding [44, 45]. POMC is synthesised in specific neurones of the ARC, which project to many brain regions, particularly elsewhere within the hypothalamus. Two melanocortin receptor subtypes (MC3-R and MC4-R) have also been located within the hypothalamus and particularly within nuclei concerned with the regulation of energy homeostasis (Fig. 2). Both of these receptors probably act

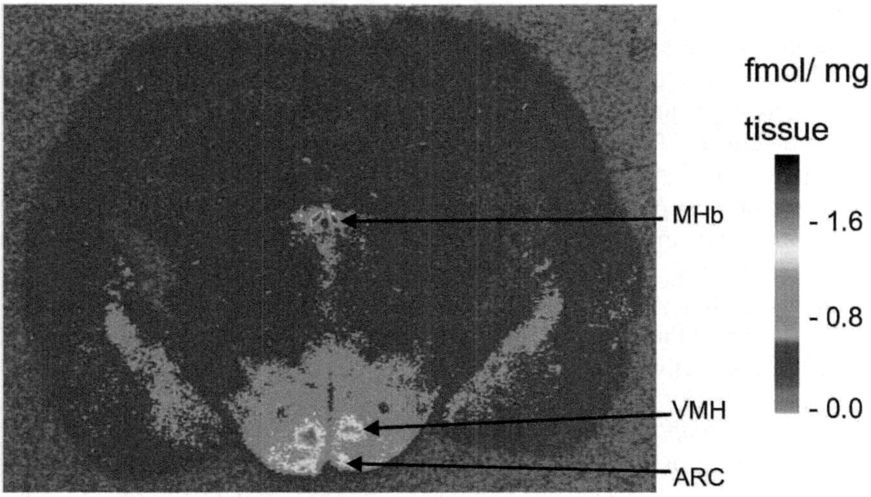

Figure 2. Distribution of MC4R binding sites in the rat hypothalamus at the levels of the arcuate nucleus (ARC) and ventromedial nucleus (VMH) of the hypothalamus. Pseudocolor image of an autoradiogram following exposure to the melanocortin ligand [^{125}I]-NDP-MSH. Key: MHb = median habenula nucleus.

to mediate the hypophagic effects of the melanocortin peptides. However, a number of studies have placed MC4-R in a central role. Huszar and colleagues reported that MC4-R null mice are hyperphagic and exhibit an obese phenotype [46]. Furthermore, highly selective MC4-R antagonists such as HS024, HS028 and SHU9119, stimulate feeding in normal and satiated rats [47, 48].

The activity of the melanocortin system is regulated by various peripheral signals, notably leptin. Approximately one third of POMC neurones express the leptin receptor and are stimulated by leptin [49]. A unique feature of the melanocortin system is the presence of an endogenous antagonist, AGRP, which confers an extra level of control on the system. Central injection of AGRP results in marked and prolonged hyperphagia and can override the leptin-induced inhibition of feeding [50].

Specific mutations affecting components of the melanocortin system, for example truncation or frameshift mutation within the POMC or MC4-R genes, are associated with rare cases of morbid obesity in humans [51, 52]. In fact, it has been suggested that such mutations could account for up to 5% of morbid obesity in children. This confirms a role for the melanocortin axis in the regulation of energy homeostasis in humans. Consequently, a vast number of MC4-R agonistic compounds have been screened as potential anti-obesity compounds. Yet, few have been clinically evaluated. However, human studies have shown that the MC4-R agonist $ACTH_{4-10}$ given intranasally can reduce hunger scores and induce modest weight loss [53]. Clearly the ability to separate such anti-obesity actions from effects on erectile activity will be a significant factor in determining whether any MC4-R agonists can be developed as anti-obesity agents.

Melanocortin antagonists are also being explored for other therapeutic applications. These block anorexia in rodents induced by central administration of exogenous α-MSH and other agonists. Recently, a number of companies, including Chiron, GSK, Amgen and Eli Lily & Co [54–57] have reported on the effects of highly-selective MC4-R antagonists and some have been shown to induce food intake *in vivo*. These compounds may prove useful in the treatment of anorectic conditions such as cancer cachexia. The evaluation of such compounds may also provide further information on routes of administration and binding characteristics which is also applicable to the development of melanocortin agonists useful for the treatment of obesity. With the intensity of research in this area it is anticipated that such MC4-R agonist will enter clinical trials within the near future.

Cocaine- and Amphetamine-Regulated Transcript (CART)

CART was initially identified as a result of its positive regulation by the psychomotor stimulants cocaine and amphetamine [58]. In the brain regions sensitive to these stimulants, CART mRNA levels were enhanced 4–5 fold. However, further investigation identified that CART mRNA was also enriched within the hypothalamus. Furthermore, obese animals were found to have little or no expression of CART and food restriction was found to reduce CART levels in the ARC of lean animals [59, 60]. Consequently, in 1998, CART was added to the list of feeding-related neuropeptides.

Current rationale places CART in the role of an endogenous satiety factor which modulates the actions of NPY and leptin. Administration of CART inhibits NPY-induced feeding [60] while central infusion of anti-CART antibodies increases food intake [60–62]. Additionally, peripheral administration of leptin to obese rodents results in increased ARC CART mRNA expression [60]. However, the precise role of CART remains difficult to determine in the light of some experimental results. For example, the peptide fragments CART(82–103) and CART(55–102) only transiently decrease NPY-induced feeding. Furthermore, CART null mice only become obese when fed a high-calorie diet [63]. Finally, it has been reported that feeding is actually increased by an intra-hypothalamic administration of CART [64]. Despite these apparent inconsistencies, CART and its receptor (which as yet remains unidentified) present another potential therapeutic target to control feeding and ultimately obesity.

Non-peptide neurotransmitters

Endocannabinoids

In the past decade, cannabinoid receptors and their putative ligands have been discovered within the CNS and linked to a number of aspects of feeding behaviour. Recently, interest has revived in the effect on appetite of the plant-derived cannabinoids and analagous molecules. Current research suggests that the endocannabinoids may be key to the hedonic aspect of eating, possibly mediating the craving for and enjoyment of the most desired (and most fattening) foods. The cannabinoid system consists of two receptors, their endogenous ligands and the uptake mechanisms and hydrolysing enzymes that regulate ligand levels. The two receptor subtypes are classified as the 'central' CB1 receptor, which is widely distributed in the CNS and many peripheral tissues, and the 'peripheral' CB2 receptor, which is not significantly expressed in the CNS [65]. It is generally accepted that the influences of cannabinoids on feeding behaviour are mediated by the CB1 receptor, which is expressed at particularly high levels in brain regions (including the hippocampus and basal ganglia) that correspond with cannabinoid-mediated behavioural effects [66, 67].

The existence of specific receptor sites indicates the presence of substances, produced within mammalian tissues, for which the cannabinoid receptors are targets. In 1992, the first endocannabinoid was isolated from porcine brain and termed anandamide from 'ananda' meaning bliss [68, 69]. Subsequent searches for additional ligands identified 2-arachidonoyl-glycerol (2-AG) [70]. Anandamide and 2-AG are considered to be the primary ligands at CB1 and CB2 receptors. However, other candidate endocannabinoids have recently been characterised including nolandin and virodhamine [71, 72]. These ligands have also been identified in various species. In conjunction with the observation that amphibian, mammalian and human CB1 receptors have a high degree of homology; this suggests that the cannabinoid signalling system plays an important physiological role. The fact that ablation of CB1 receptors results in mice with a lean phenotype, resistance to dietary-induced obesity and enhanced leptin sensitivity [73] suggests that they represent an orexigenic component of the energy homeostatic circuitry.

Both exogenous and endogenous (anadamide and 2-AG) cannabinoids stimulate feeding [74–76]. The hyperphagia is powerful; peripheral administration of the exogenous cannabinoid Δ^9-tetrahydrocannabinol (Δ^9-THC) stimulates feeding as potently as does central administration of NPY [77]. However, targeted disruption of the gene encoding NPY has little influence on energy balance [78] suggesting that the NPY system is a less critical player than the cannabinoid system in the chronic maintenance of energy balance.

As the orexigenic effect of the cannabinoid agonists is blocked by the CB1 specific antagonist SR 141716, but not by an antagonist of CB2 receptors, this suggests that these actions on feeding are mediated by the central CB1 receptor. Furthermore, as administration of SR 141716 alone suppresses food intake in rodents [79, 80] this suggests that tonic endocannabinoid activity at these receptors may be a key component of appetite regulation. This tonic activity is further supported by direct measurements of brain endocannabinoid levels in response to fasting and feeding. Fasting increases levels of anandamide and 2-AG in the nucleus accumbens, and to a lesser extent in the hypothalamus, whereas 2-AG levels decline in the hypothalamus with feeding [81]. However, levels in the cerebellum, a region not directly involved in the control of feeding, are not influenced by nutritional status [81].

The mechanisms of cannabinoid-induced hyperphagia remain to be elucidated. There is a body of evidence that points towards involvement of established homeostatic pathways, many of which are regulated by the hormone leptin and operate within hypothalamic nuclei. Firstly, leptin administration decreases hypothalamic levels of anandamide and 2-AG, while endocannabinoid levels in the cerebellum are unaffected [82]. In addition, anandamide increases Fos expression in the PVN of rodents [83, 84] while administration into the VMH of satiated rats induces significant hyperphagia [85]. Furthermore, defective leptin signalling in *ob/ob* and *db/db* mice and *fa/fa* Zucker rats is associated with elevated levels of hypothalamic endocannabinoids and these are reduced in *ob/ob* mice following leptin treatment [82]. Moreover, CB1 is expressed in a number of leptin-regulated key hypothalamic peptidergic systems of appetite regulation, including those producing CART in the ARC, and melanin-concentrating hormone and orexin in the LHA [86]. Finally, evidence has been obtained which indicates an interaction between CB1 receptors and the melanocortin system, with the observation that subanorectic doses of SR 141716 and the melanocortin receptor agonist α-MSH synergistically attenuate baseline feeding when combined [87].

However, hypothalamic 2-AG levels have been found to increase with food deprivation and decline with feeding [81], suggesting that once initiated, eating no-longer depends on hypothalamic endocannabinoids for maintenance. Furthermore, CB1 receptor binding in the hypothalamus of dietary obese rats is low and unaltered, and no relationship has been identified between CB1 receptor binding density (in any brain region) and leptin levels in these animals [67]. These results are notable, drawing attention away from the hypothalamus and leptin-regulated pathways, and arguing against a role for hypothalamic cannabinoids in influencing food intake in dietary obesity.

In contrast, significant reductions in CB1 receptor binding density, consistent with increased receptor activity, have been identified in the forebrain and hippocampus following 10 weeks of palatable diet feeding (Fig. 3) [67]. These brain areas are either directly involved in the hedonic aspects of eating or are connected to reward-related brain areas [88–90]. Association of the cannabinoid system with reward processes is supported by a number of other lines of evidence. SR 141716 antagonises the hunger induced by anandamide and 2-AG. However, it also produces changes in ingestive behaviour when administered alone. SR 141716 selectively inhibits consumption of palatable food and drink, with decreased intakes of sucrose, alcohol and a sweet diet observed in rats, mice and marmosets respectively [91, 92]. These results suggest that the central cannabinoid system may act to amplify reward indices. In addition, the cannabinoid system appears to interact with known opioidergic reward pathways, indicated by the synergistic actions of SR 141716 or the cannabinoid inverse agonist AM251 with the opioid receptor antagonists, naloxone and nalmefene, on food intake [93, 94]. Furthermore, evidence in humans supports cannabinoid involvement in food reward, with hyperphagic effects of marijuana in human volunteers being principally attributed to an

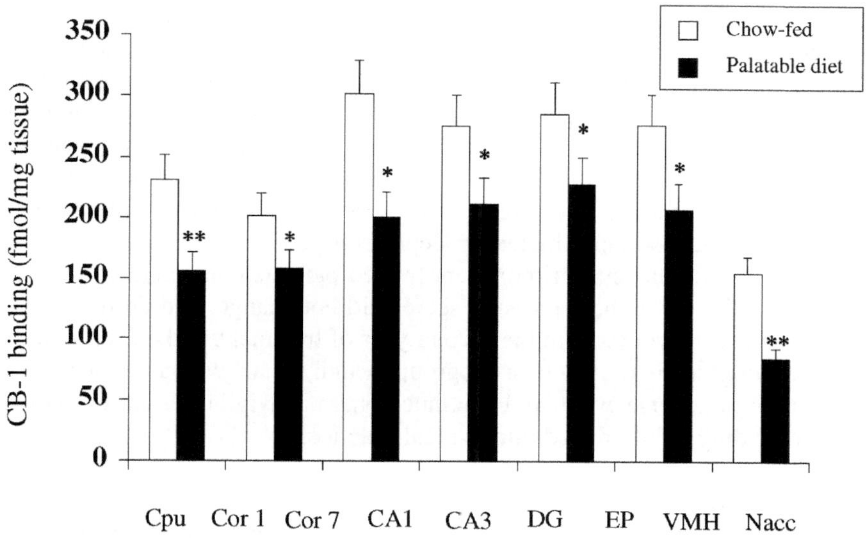

Figure 3. Regional cannabinoid receptor density in rats given palatable diet for 10 weeks (n = 8) and in chow-fed controls (n = 8). Data are mean ± SEM * $P < 0.05$ *versus* controls. Other brain regions examined showed no significant effects of palatable-diet feeding (data not shown).

increase in the consumption of highly palatable sweet foods such as chocolate and biscuits [95].

In stark contrast to the cannabinoid system, leptin regulated hypothalamic orexigenic neuropeptide systems, such as NPY, are reported to be switched off under conditions of excess intake [22]. Additionally, simultaneous deletion of the two most potent orexigenic neuropeptides known to date – NPY and AGRP – fails to produce a lean phenotype, demonstrating the apparent redundancy of neuroendocrine factors that drive food intake. However, none of the appetite-stimulating peptides identified to date are able to compensate for the lack of endogenous cannabinoid action, reflecting its crucial role in the regulation of energy balance [86]. Therefore, the cannabinoid system appears to be a logical target for the treatment of lifestyle related obesity in human subjects.

Promising preclinical findings have been confirmed in human studies. Recent Phase III clinical trials with the experimental drug rimonabant (Acomplia®; SR 141716 – a CB1 specific antagonist) have indicated that the drug can effectively reduce weight and adiposity in obese people (for a critical evaluation of recently reported clinical data see [96]). The first peer-reviewed reports of Phase III trials have recently been published. RIO-Europe and RIO-North America evaluate the efficacy of rimonabant in obese or overweight patients without comorbidities [97, 98], while RIO-lipids assesses its use in patients with untreated dyslipidaemia [99] and RIO-Diabetes in patients treated with metformin or sulfonylurea [100]. The reports show that the rimonabant-treated groups consistently show weight loss matching or exceeding the effects obtained with earlier classes of anti-obesity drugs. Additionally, the metabolic effects of rimonabant seen in animal models were replicated in the clinical trial. Rimonabant significantly increased high density lipoproteins, lowered plasma free fatty acid levels, countered insulin resistance and reduced the prevalence of metabolic syndrome. Interestingly, the improvements in almost half of the metabolic risk factors were greater than could be attributed to weight loss alone suggesting a direct peripheral effect of rimonabant. However, mood disorders, including depression and anxiety, were reported with a greater frequency in rimonabant-treated patients compared to controls in these studies. Overall, depression scores did not change, and dropouts due to adverse events were low in the second year of treatment in the 2 year studies. This suggests there may be a subgroup of individuals who are susceptible to this type of adverse event, and continued pharmacovigilance will be necessary as the drug is introduced into clinical practice.

Serotonin (5-HT)

It is now well established that 5-HT acts to inhibit feeding and stimulate thermogenesis in rodents [101]. 5-HT neurones are located in the raphe nuclei of the midbrain, but project extensively throughout the CNS, particularly to the hypothalamic appetite-regulating nuclei. These neurones express leptin receptors and may partly mediate the effects of leptin on energy homeostasis [102].

Although links between the 5-HT system and the regulation of feeding were determined in the 1970s, it is only relatively recently that the receptor subtypes responsible for 5-HT mediated hypophagia have been established. Over 15 5-HT receptor subtypes have been identified to date. These contribute to the regulation of feeding to different degrees. Research into potential treatments for obesity has centred upon the 5-HT_{1A}, 5-HT_{1B}, 5-HT_{2A} and 5-HT_{2C} subtypes. Of these, the 5-HT_{2C} subtype appears to play a particularly important role in regulating food intake. Knockout mice possessing no functional 5-HT_{2C} receptors demonstrate marked hyperphagia and consequently obesity [103]. Furthermore, 5-HT_{2C} specific antagonists inhibit the hypophagic actions of sibutramine and other 5-HT enhancing drugs.

While most of the focus on 5-HT and weight control has been on drugs agonising the 5-HT_{2C} receptor, other targets in the 5-HT system exist. The 5-HT_6 receptor is one of the most recent additions to the 5-HT receptor family. It is almost exclusively expressed within the CNS with high levels in cortical and limbic regions [104]. The recent development of specific 5-HT_6 ligands has indicated potential roles for this receptor in a number of physiological processes, including feeding [105]. 5-HT_6 selective antagonists have been reported to produce significant reductions in food intake when administered to *ob/ob* mice, with these hypophagic effects being accompanied by significant reductions in body weight and insulin levels [106]. Such results identify 5-HT_6 receptor antagonists as potential anti-obesity agents.

How does the hypothalamus sense energy requirements?

This is one of the most crucial questions in the understanding of energy homoeostasis. Numerous neural pathways have been identified, including those transmitting sensory information from the upper gastrointestinal tract, abdominal viscera and oral cavity (taste). These inputs are all relayed to and integrated within the NTS and the parabrachial nucleus (the 'pontine taste centre'). Signals initiated by mechanical and chemical stimulation of the stomach and intestine and neural inputs from glucoceptors in the liver, that respond to falls in the portal venous glucose concentration [107], also converge on the NTS via the vagal nerve. Afferent fibres then relay these signals to the hypothalamus and other appetite-regulating areas of the brain. Various circulating hormones have also been proposed to signal energy status to the CNS. Leptin and insulin are postulated to be 'adiposity signals' for the long-term regulation of body weight by the brain.

Leptin

Leptin is the 146-amino acid residue product of the *ob* gene. It is produced primarily by white adipose tissue and at lower levels in brown adipose tissue (in rodents), stomach, skeletal muscle, the placenta and mammary glands [108–110]. Leptin is highly conserved among mammalian species: human leptin shares 84% and 83% sequence homology with the mouse and rat peptides respectively [111]. It is now generally accepted that leptin acts as a signal to indicate the size of adipose tissue stores to the CNS and in doing so acts as an afferent signal in a negative feedback loop in which its anti-obesity actions tend to counteract any increase in fat mass [112]. Thus, leptin is a key component of the physiological systems that regulate food intake and body weight.

Considerable evidence for an involvement for leptin in the central regulation of feeding has been provided by the study of three genetic models of obesity, *ob/ob* and *db/db* mice and the *fa/fa* rat [113, 114]. All of these models have single gene mutations affecting components of the leptin signalling system, that consequently result in severe obesity due to hyperphagia. Consistent with these observations, acute central or peripheral administration of leptin to rodents reduces food intake, while chronic administration also decreases fat mass without altering lean tissue content [115–117].

Leptin exerts these influences on energy homeostasis via its actions at various neural targets located in the hypothalamus, medulla and other sites which express the extended functional variant of the leptin receptor (OB-Rb) [118]. Peripherally administered leptin enters the mediobasal hypothalamus and ARC, where the BBB is modified to allow the passage of large molecules [119]. Within the ARC leptin's targets include appetite-stimulating neurones (e.g., those expressing NPY) that are inhibited by leptin, and appetite-inhibiting neurones (e.g., those expressing POMC) that are stimulated by leptin [120, 121].

The convincing role of leptin in regulating body weight in rodents raised scientific, medical and commercial expectations that human obesity would be due to leptin deficiency. However, single-gene defects in the genes encoding leptin or its receptors are now known to cause obesity in only a few rare cases in humans. Obese people and rodents conform to the general rule that plasma leptin concentrations increase with increasing fat mass, and as such demonstrate hyperleptinaemia. The counter-regulatory effects of leptin which act to decrease food intake (via the hypothalamus) in normal weight individuals are therefore evidently lacking in the obese. This apparent state of leptin insensitivity or resistance is proposed to be a result of defective leptin entry into the CNS [122].

Leptin resistance may also explain disappointing results to some recent trials of peripherally administered exogenous leptin or its analogues. The results of a double-blind, placebo-controlled, dose-escalating trial using R-muHu-Leptin suggest that high leptin dosages can reduce body fat and implies that leptin resistance can be overcome with sufficiently high leptin concentrations. However, the magnitude of weight loss observed was not dramatic and did not exceed that observed with conventional anti-obesity agents [123]. A further problem with this trial was the route of administration – subcutaneous injection, resulting in erythema at the injection site. However, attempts to improve formulations or modes of delivery, in order to prolong biological action, have also yielded disappointing results. For example, pegylated leptin (PEG-OB), which has an extended half life (>48 h), has been shown to be effective in decreasing food intake and body weight in rats treated for 8 days [124, 125]. However, a 12 week study in obese humans, given in association with a hypocaloric diet, showed no significant loss of body weight or fat when compared to placebo [126].

Circumvention of resistance to leptin may potentially be achieved by activation or potentiation of the leptin related signalling pathway. Ciliary neurotropic factor (CNTF) is one molecule apparently capable of achieving this. CNTF was discovered to induce substantial weight loss during its initial trials for treating patients with motor neurone disease (amyotrophic lateral sclerosis, CNTF study, 1996). It has since been found that CNTF and its receptor complex are closely associated with the leptin receptor in the hypothalamus. Additionally, they appear to coactivate the STAT3 transcription factor [127] and inhibit NPY-ergic signalling [128]. Moreover, administration of CNTF to wild type rodents reduces food intake, resulting in weight loss which primarily occurs due to loss of fat mass. Interestingly, these effects are maintained long after treatment has been ceased [128–130]. Furthermore, CNTF appears to be effective at inducing weight loss in a variety of rodent models of obesity, including *ob/ob* mice [129], *db/db* mice [129], MC4-R deficient A^y mice [131] and in dietary-induced obese wild type mice [130]. Overall, CNTF induces significant and pro-longer weight and fat loss, without rebound weight gain, even in situations where leptin has previously been found to be ineffective [132].

Genetically re-engineered CNTF (Axokine) is under active development by Regeneron Pharmaceuticals. Data from Phase II trials demonstrated that morbidly obese subjects treated with CNTF for 12 weeks lost 5 kg more weight than placebo-treated controls. Most impressively, there was no rebound weight gain observed during a 36 week follow up period after treatment was stopped (Regeneron Pharmaceuticals, 2001). However, it remains to be seen whether CNTF can be used as an effective therapy as two thirds of the individuals who took Axokine in the trials developed neutralising antibodies [133].

This uncertain success of leptin therapy has generated much interest in the development of leptin-related peptide agonists [134–136]. Discrete bioactive domains of leptin have been identified. Of particular interest is the leptin fragment leptin(116–130), which appears to mediate its effects (reduced food intake, body weight gain and glucose levels) independent of the leptin receptor in *db/db* mice [137], suggesting a potential means of resolving leptin resistance. Another approach to this problem is the targeting of other components of the leptin signalling pathway. For example, antisense oligonucleotides against the product of gene 46a (involved in leptin breakdown) have been shown to cause a reduction in 24 h food intake in *ob/ob* mice greater than that of leptin treatment [138].

Insulin

Insulin, like leptin, is an anti-obesity signal which acts centrally to modify energy balance and metabolism. When injected centrally to rodents and monkeys insulin inhibits feeding and stimulates thermogenesis [139, 140], while injection of anti-insulin antibodies into the VMH stimulates feeding [141]. Generally, insulin circulates at levels broadly proportional to fat mass and as such concentrations are decreased in all states of energy deficit such as starvation, lactation, physical exercise and insulin-deficient diabetes. Consistent with its role in the regulation of energy balance, such hypoinsulinaemic states are associated with hyperphagia, while central administration of insulin normalises food intake. These effects of insulin are mediated in part by its inhibition of hypothalamic NPY neurones [142, 143].

Insulin crosses the BBB to enter the brain via a specialised transport mechanism dependent on insulin receptors expressed by brain microvessels. Insulin receptors are also widely distributed throughout the brain with a particularly dense expression within the hypothalamus (notably the ARC and PVN) [144]. Selective knockout of the brain insulin receptors leads to hyperphagia and obesity [145] confirming insulin's role in the regulation of food intake and body weight.

Small non-peptide molecules have been generated which mimic the physiological actions of insulin in terms of the regulation of energy balance. Cpd 1 and Cpd2 are two such mimetics. Cpd 1 has been reported to produce a dose-dependent reduction of food intake and body weight in normal animals and to stop the progress of obesity in diet-induced obese mice in conjunction with lessened adiposity and insulin resistance [146]. These results support the use of these insulin-mimetics as potential anti-obesity agents.

Other peripheral signals

Cholecystokinin (CCK)

CCK, the archetypal satiety peptide, is released from cells lining the gut upon nutrient stimulation [147]. It dose-dependently decreases meal size in rodents without altering body weight. Similarly, intravenous infusion of CCK_8 also decreases hunger and feeding in humans, but again does not influence weight [148].

The actions of CCK involve both peripheral and central pathways and are mediated through two different receptor subtypes. In the periphery the effects of CCK are mediated by CCK1 receptors (formerly CCK_A). These are located on the ends of vagal sensory fibres that extend from the gut to the NTS in the brain. Vagotomy blocks the hypophagic actions of peripherally administered CCK, highlighting the importance of this pathway. Both leptin [149] and insulin [150] enhance the satiation effects of CCK, with leptin being shown to potentiate the activation of NTS neurones by CCK [151, 152]. Other signals of energy homeostasis may also module the activity of NTS neurones. NPY neurones, MC4-R and POMC neurones are all located in the NTS. Furthermore, activation of MC4-R is required for CCK to suppress feeding and POMC neurones have been shown to be activated by CCK [153]. By contrast, central administration of CCK inhibits food intake probably via activation of CCK2 (formerly CCK_B) receptors located in the VMH and PVN [154]. The importance of these receptors in the regulation of normal food intake is uncertain. Dourish and colleagues suggest that endogenous CCK acts predominantly at central CCK2 receptors. However, the importance of CCK1 receptors has also been confirmed by the observation that genetic ablation of these receptors in OLEFT rats leads to hyperphagia and obesity [155].

In the late 1990s several potent, selective and orally active CCK agonists entered clinical trials for obesity. However, to date, only one – GI-181771 (GSK) – has progressed into Phase II clinical trials, but not as far as a Phase III clinical development programme.

Ghrelin

Ghrelin is a recently discovered gastric peptide, which is predominately secreted from the stomach and is the only known endogenous ligand for the growth hormone (GH) secretagogue receptor (SR). As well as stimulating GH release ghrelin stimulates feeding and weight gain in rodents [156–158]. Importantly, weight gain appears to be due to increased fat mass, with no changes in lean mass or longitudinal growth observed. These orexigenic and adipogenic effects are mediated, at least in part, by stimulating ARC NPY/AGRP neurones; administration of antibodies or antagonists to these neuropeptides abolishes ghrelin action [157, 159].

In humans ghrelin enhances appetite, increases feelings of hunger and increases food intake following intravenous administration [160]. In lean humans ghrelin levels rise before meals and fall in response to feeding [161–163]. Consistent with this, plasma ghrelin levels rise in anorexic individuals and are suppressed in the obese [164]. However, the post-prandial suppression of ghrelin appears to be absent from obese individuals and this may contribute to the development of obesity [165]. Furthermore, polymorphisms of the ghrelin gene may present a genetic predisposition to obesity. Indeed, significantly increased ghrelin levels are observed in morbidly obese patients with Prader-Willi syndrome [166].

In theory, ghrelin antagonists could be used to treat obesity. One, [D-Lys^3]GHRP-6 has been shown to reduce ghrelin-stimulated food intake in mice [167]. However, as ghrelin levels are already lowered in obese individuals they may offer little benefit in trying to lower ghrelin further. Additionally, it may prove difficult to block ghrelin's orexigenic actions without blockade of its other effects; impaired GH secretion may lead to undesirable side effects including elevated cardiac risk [168]. However, new approaches are being employed in an attempt to target ghrelin's orexigenic actions. Spiegelmers, stable RNA-based compounds, have been described, which inhibit ghrelin mediated GHSR activation and GH release in rodents [169]. It has also been suggested that a receptor other than GHSR mediates the feeding effects of ghrelin [170, 171]. Identification and targeting of such may offer a more viable pharmacological approach to reduce food intake and body weight.

PYY$_{(3-36)}$

PYY$_{(3-36)}$ is a gut peptide belonging to the same peptide family as NPY. It is generated by proteolytic cleavage of peptide YY (PYY) and is released from the ileum and colon in response to feeding in proportion to the calorie content of the meal [172]. Similar to CCK, PYY$_{(3-36)}$ is a satiety signal which only transiently inhibits food intake. However, unlike CCK it does not achieve these effects via neuronal activation in the brainstem, but directly in hypothalamic neurones that express POMC [173]. The feeding effects of PYY$_{(3-36)}$ are proposed to occur as a consequence of agonistic activity at NPY Y_2 receptors. The autoreceptors are located presynaptically on NPY/AGRP neurones in the ARC and their activation blocks the release of these orexigenic peptides. Consequently, the tonic inhibitory tone normally exerted on the appetite-suppressing POMC/CART neurones is removed [174]. In support of this, Batterham and colleagues demonstrated that peripheral or central administration of PYY$_{(3-36)}$ is able to reduce food intake and body weight in mice, but that this effect is absent from Y_2 null mice [174].

PYY$_{(3-36)}$ has also been shown to acutely reduce appetite and food intake in humans [175]. The observation that fasting and post-prandial levels of the peptide are reduced in obese individuals further supports PYY$_{(3-36)}$ as a credible target for the pharmacological control of obesity. Initial results of Phase 1 trials using a PYY nasal spray are promising, with an average 8.2% reduction in food intake (Nastech Pharmaceuticals Co Inc). However, evidence of its ability to suppress feeding and body weight long-term is required.

Concluding remarks

One of the major barriers to understanding the mechanisms associated with the development of obesity is the necessity of extrapolating experimental data from rodents to humans. However, it is now clear that some of the lessons learnt from experimental mammals, such as lesions of the VMH or LHA demonstrating the importance of the hypothalamus to the control of food intake, are also applicable to man.

Unravelling the myriad of interacting systems controlling both episodic (short-term) and tonic (long-term) signals that converge on the hypothalamus offers a further obstacle to the treatment of obesity. Due to the numerous pathways involved in the regulation of body weight it is likely that combination therapies will prove necessary to achieve long-term weight loss goals. When targeted independently, these up-and-coming drug targets are likely to offer effective weight loss treatment only for specific patient groups.

References

1 Kristensen P, Judge ME, Thim L, Ribel U, Christjansen KN, Wulff BS, Clausen JT, Jensen PB, Madsen OD, Vrang N et al. (1998) Hypothalamic CART is a new anorectic peptide regulated by leptin. *Nature* 393: 72–76
2 Elias CF, Lee C, Kelly J, Aschkenasi C, Ahima RS, Couceyro PR, Kuhar MJ, Saper CB, Elmquist JK (1998) Leptin activates hypothalamic CART neurons projecting to the spinal cord. *Neuron* 21: 1375–1378
3 Elmquist JK, Bjorbaek C, Ahima RS, Flier JS, Saper CB (1998) Distribution of leptin receptor mRNA isoforms in the rat brain. *J Comp Neurol* 395: 535–547
4 Elmquist JK, Elias C, Saper C (1999) From lesions to leptin: hypothalamic control of food intake and body weight. *Neuron* 22: 221–232
5 Ono T, Nishino H, Sasaka K, Muramoto K, Yano I, Simpson A (1978) Paraventricular nucleus connections to the spinal cord and pituitary. *Neurosci Lett* 10: 141–146
6 Shor-Posner G, Azar AP, Insinga S, Leibowitz SF (1986) Deficits in the control of food intake after paraventricular nucleus lesions. *Physiol Behav* 35: 883–890
7 Stellar E (1954) The physiology of motivation. *Psychol Rev* 61: 5
8 Meister B, Ceccatelli S, Hokfelt T, Anden NE, Theodorsson E (1989) Neurotransmitters, neuropeptides and binding sites in the rat mediobasal hypothalamus: effects of monosodium glutamate (MSG) lesions. *Exp Brain Res* 76: 343–368
9 Broberger C, Johansen J, Johasson C, Schalling M, Hokfelt T (1998) The neuropeptide Y/agouti gene-related protein (AGRP) brain circuitry in normal, anorectic and monosodium glutamate-treated mice. *Proc Natl Acad Sci* 95: 15043–15048
10 Horvath TL, Diano S, van den Pol AN (1999) Synaptic interaction between hypocretin (orexin) and neuropeptide Y cells in the rodent and primate hypothalamus: a novel circuit implicated in metabolic and endocrine regulations. *J Neurosci* 19: 1072–1087
11 Hu YH, Bloomquist BT, Cornfield LJ, DeCarr LB, Flores-Riveros JR, Friedman L, Jiang PL,

Lewis-Higgins L, Sadlowski Y, Schaefer J et al. (1996) Identification of a novel hypothalamic neuropeptide Y receptor associated with feeding behavior. *J Biol Chem* 271: 26315–26319
12 Bernardis LL, Bellinger LL (1996) The lateral hypothalamic area revisited: Ingestion behavior. *Neurosci Biobehavior Rev* 20: 189–287
13 Kalra SP, Dube MG, Pu SY, Xu B, Horvath TL, Kalra PS (1999) Interacting appetite-regulating pathways in the hypothalamic regulation of body weight. *Endocr Rev* 20: 68–100
14 Mountjoy K, Mortrud M, Low M, Simerly R, Cone R (1994) Localization of the melanocortin-4 receptor (MC4-R) in neuroendocrine and autonomic control circuits in the brain. *Mol Endocrinol* 8: 1298–1308
15 Grill H, Ginsberg A, Seeley R, Kaplan J (1998) Brainstem application of melanocortin receptor ligands produces long-lasting effects on feeding and body weight. *J Neurosci* 18: 10128–10135
16 Hyde TM, Miselis RR (1983) Effects of area postrema caudal medial nucleus of solitary tract lesions on food intake and body weight. *Am J Phsyiol* 244: R577–R587
17 Allen YS, Adrian TE, Allen JM, Tatemoto K, Crow TJ, Bloom SR, Polak JM (1983) Neuropeptide Y distribution in rat brain. *Sci* 221: 877–879
18 Stanley BG, Daniel DR, Chin AS, Leibowitz SF (1985) Paraventricular nucleus injections of peptide YY and neuropeptide Y preferentially enhance carbohydrate ingestion. *Peptides* 6: 1205–1211
19 Turnbull AV, Ellershaw L, Masters DJ, Birtles S, Boyer S, Carroll D, Clarkson P, Loxham SJ, McAulay P, Teague JL et al. (2002) Selective antagonism of the NPY Y5 receptor does not have a major effect on feeding in rats. *Diabetes* 51: 2441–2449
20 Beck-Sickinger AG, Jung G (1995) Structure-activity relationships of neuropeptide Y analogues with respect to Y1 and Y2 receptors. *Biopolymers* 37: 123–142
21 Wang J, Leibowitz KL (1997) Central insulin inhibits galanin and neuropeptide Y gene expression and peptide release in intact rats. *Brain Res* 777: 231–236
22 Widdowson PS, Upton R, Henderson L, Buckingham R, Wilson S, Williams G (1997) Reciprocal regional changes in brain NPY receptor density during dietary restriction and dietary-induced obesity in the rat. *Brain Res* 774: 1–10
23 Kawauchi H, Kawazoe I, Tsubokawa M, Kishida M, Baker BI (1983) Characterization of melanin-concentrating hormone in chum salmon pituitaries. *Nature* 305: 321–323
24 Presse F, Nahon JL, Fischer WH, Vale W (1990) Structure of the human melanin concentrating hormone mRNA. *Mol Endocrinol* 4: 632–637
25 Vaughan JM, Fischer WH, Hoeger C, Rivier J, Vale W (1989) Characterization of melanin-concentrating hormone from rat hypothalamus. *Endocrinology* 125: 1660–1665
26 Naito N, Kawazoe I, Nakai Y, Kawauchi H (1988) Melanin-concentrating hormone-like immunoreactive material in the rat hypothalamus; charcterization and subcellular localization. *Cell Tissue Res* 253: 291–295
27 Skofitsch G, Jacobowitz DM, Zamir N (1985) Immunohistochemical localization of melanin-concentrating hormone-like peptide in the rat brain. *Brain Res Bull* 15: 635–649
28 Elias CF, Lee CE, Kelly JF, Ahima RS, Kuhar M, Saper CB, Elmquist JK (2001) Characterization of CART neurons in the rat and human hypothalamus. *J Comp Neurol* 432: 1–19
29 Broberger C (1999) Hypothalamic cocaine- and amphetamin-regulated transcript (CART) neurons: histochemical relationship to thyrotropin-releasing hormone, melanin-concentrating hormone, orexin/hypocretin and neuropeptide Y. *Brain Res* 848: 101–113
30 Tan CP, Sano H, Iwaasa H, Pan J, Sailer AW, Hreniuk DL, Feighner SD, Palyha OC, Pong SS, Figueroa DJ et al. (2002) Melanin-concentrating hormone receptor subtypes 1 and 2: species-specific gene expression. *Genomics* 79 : 785–792
31 Chambers J, Ames RS, Bergsma D, Muir A, Fitzgerald LR, Hervieu G, Dytko GM, Foley JJ, Martin J, Liu WS et al. (1999) Melanin-concentrating hormone is the cognate ligand for the orphan G-protein-coupled receptor SLC-1. *Nature* 400: 261–265
32 Hervieu GJ, Cluderay JE, Harrison D, Meakin J, Maycox P, Nasir S, Leslie RA (2000) The distribution of the mRNA and protein products of the melanin-concentrating hormone (MCH) receptor gene, slc-1, in the central nervous system of the rat. *Eur J Neurosci* 12: 1191–1216
33 Tsukamura H, Thompson RC, Tsukahara S, Ohkura S, Maekawa F, Morivama R, Niwa Y, Foster DL, Maeda K (2000) Intracerebroventricular administration of melanin-concentrating hormone suppresses pulsatile luteinising hormone release in the female rat. *J Neuroendocrinol* 12: 529–534
34 An S, Cutler G, Zhao JJ, Huang SG, Tian H, Li W, Liang L, Rich M, Bakleh A, Du J et al. (2001) Identification and characterization of a melanin-concentrating hormone receptor. *Proc Natl Acad Sci USA* 98: 7576–7581

35 Qu D, Ludwig DS, Gammeltoft S, Piper M, Pelleymounter MA, Cullen MJ, Mathes WF, Przypek R, Kanarek R, Maratos-Flier E (1996) A role for melanin-concentrating hormone in the central regulation of feeding behaviour. *Nature* 380: 243–247
36 Rossi M, Choi SJ, O'Shea D, Miyoshi T, Ghatei MA, Bloom SR (1997) Melanin-concentrating hormone acutely stimulates feeding, but chronic administration has no effect on body weight. *Endocrinol* 138: 351–355
37 Ludwig DS, Tritos NA, Mastaitis JW, Kulkarni R, Kokkotou E, Elmquist J, Lowell B, Flier JS, Maratos-Flier E (2001) Melanin-concentrating hormone overexpression in transgenic mice leads to obesity and insulin resistance. *J Clin Invest* 107: 379–386
38 Shimada M, Tritos NA, Lowell BB, Flier JS, Maratos-Flier E (1998) Mice lacking melanin-concentrating hormone are hypophagic and lean. *Nature* 396: 670–673
39 Segal-Lierberman G, Bradley RL, Kokkotou E, Carlson M, Trombly DJ, Wang X, Bates S, Myers MG Jr, Flier JS, Maratos-Flier E (2003) Melanin-concentrating hormone is a critical mediator of the leptin-deficient phenotype. *Proc Natl Acad Sci USA* 100: 10085–10090
40 Chen Y, Hu C, Hsu CK, Zhang Q, Bi C, Asnicar M, Hsiung HM, Fox N, Slieker LJ, Yang DD et al. (2002) Targeted disruption of the melanin-concentrating hormone receptor-1 results in hyperphagia and resistance to diet-induced obesity. *Endocrinology* 143: 2469–2477
41 Marsh DJ, Weingarth DT, Nove DE, Chen HY, Trumbauer ME, Chen As, Guan XM, Jiang MM, Feng Y, Camacho RE et al. (2002) Melanin-concentrating hormone 1 receptor-deficient mice are lean, hyperactive, and hyperphagic and have altered metabolism. *Proc Natl Acad Sci USA* 99: 3240–3245
42 Takekawa S, Asami A, Ishihara Y, Terauchi J, Kato K, Shimomura Y, Mori M, Murakoshi H, Kato K, Suzuki N et al. (2002) T-226296: a novel, orally active and selective melanin-concentrating hormone receptor antagonist. *Eur J Pharmacol* 438: 129–135
43 Borowsky B, Durkin MM, Ogozalek K, Marzabadi MR, DeLeon J, Lagu B, Heurich R, Lichtblau H, Shaposhnik Z, Daniewska I et al. (2002) Antidepressant, anxiolytic and anorectic effects of a melanin-concentrating hormone-1 receptor antagonist. *Nat Med* 8: 825–830
44 Fan W, Boston BA, Kesterson RA, Hruby VJ, Cone RD (1997) Role of melanocortinergic neurons in feeding and the agouti obesity syndrome. *Nature* 385: 119–120
45 Abbott CR, Rossi M, Kim M, Al Ahmed SH, Taylor GM, Ghatei MA, Smith DM, Bloom SR (2000) Investigation of the melanocyte stimulating hormones on food intake. Lack of evidence to support a role for the melanocortin-3-receptor. *Brain Res* 869: 203–210
46 Huszar D, Lynch CA, Fairchild-Huntress V, Dunmore JH, Fang Q, Berkemeier LR, Gu W, Kesterson RA, Boston BA, Cone RD et al. (1997) Targeted disruption of the melanocortin-4 receptor results in obesity in mice. *Cell* 88: 131–141
47 Schioth HB, Mucenience R, Mutulis F, Bouifrouri AA, Mutule I, Wikberg JES (1999) Further pharmacological characterisation of the selective melanocortin 4 receptor antagonist HSO14: Comparison with SHU9119. *Neuropeptides* 33: 191–196
48 Benoit SC, Schwartz MW, Lachey JL, Hagan MM, Rushing PA, Blake KA, Yagaloff KA, Kurylko G, Franco L, Danhoo W et al. (2000) A novel selective melanocortin-4 receptor agonist reduces food intake in rats and mice without producing aversion consequences. *J Neurosci* 20: 3442–3448
49 Cheung CC, Clifton Dk, Steiner RA (1997) Proopiomelanocortin neurons are direct targets for leptin in the hypothalamus. *Endocinol* 138: 4489–4492
50 Ebihara K, Ogawa Y, Katsuura G, Numata Y, Masuzaki H, Satoh N, Tamaki M, Yoshioka T, Hayase M, Matsuoka N et al. (1999) Involvement of agouti-related protein, an endogenous antagonist at the hypothalamic melanocortin receptor; in leptin action. *Diabetes* 48: 2028–2033
51 Krude H, Biebermann H, Luck W, Horn R, Brabant G, Gruters A (1998) Severe early-onset obesity, adrenal insufficiency and read hair pigmentation caused by POMC mutations in humans. *Nat Genet* 19: 155–157
52 Yeo GSH, Farooqi S, Aminian S, Halsall DJ, Stanhope RG, O'Rahilly S (1998) A frameshift mutation in MC4-R associated with dominantly inherited human obesity. *Nat Genet* 20: 111–112
53 Krude H, Bieberman H, Schnabel D, Tansek MZ, Theunissen P, Mullis PE, Gruters A (2003) Obesity due to proopiomealnocortin deficiency: three new cases and treatment trials with thyroid hormone and ACTH4–10. *J Clin Endocrinol Metab* 88: 4633–4640
54 Koikov LN, Ebetino FH, Solinsky MG, Cross-Doersen D, Knittel JJ (2003) Sub-nanomolar hMC1R agonists by end-capping of the melanocortin tetrapeptide His-D-Phe-Arg-Trp-NH(2). *Bioorg Med Chem Lett* 13: 2647–2650

55 Kulesza A, Ebetino FH, Mishra RK, Cross-Doersen D, Mazur AW (2003) Synthesis of 2,4,5-trisubstituted tetrahydropyrans as peptidomimetic scaffolds for melanocortin receptor ligands. *Org Lett* 5: 1163–1166
56 Xi N, Hale C, Kelly MG, Norman MH, Stec M, Xu S, Baumgartner JW, Fotsch C (2004) Synthesis of novel melanocortin 4 receptor agonists and antagonists containing a succinamide core. *Bioor Med Chem Lett* 14: 377–381
57 Richardson TI, Ornstein PL, Briner K, Fisher MJ, Backer RT, Biggers CK, Clay MP, Emmerson PJ, Hertel LW, Hsiung HM et al. (2004) Synthesis and structure-activity relationships of novel arylpiperazines as potent and selective agonists of the melanocortin subtype 4 receptor. *J Med Chem* 47: 744–755
58 Kuhar MJ, Dall Vechia SE (1999) CART peptides: novel addiction- and feeding-related neuropeptides. *Trends Neurosci* 22: 316–320
59 Douglass J, Daoud S (1996) Characterization of the human cDNA and genomic DNA encoding CART: a cocaine- and amphetamine-regulated transcript. *Gene* 169: 241–245
60 Kristensen P, Judge ME, Thim L, Ribel U, Christjansen KN, Wulff BS, Clausen JT, Jensen PB, Madsen OD, Vrang N et al. (1998) Hypothalamic CART is a new anorectic peptide regulated by leptin. *Nature* 393: 72–76
61 Thim L, Kristensen P, Larsen PJ, Wulff BS (1998) CART, a new anorectic peptide. *Int J Biochem Cell Biol* 30: 1281–1284
62 Thim L, Nielsen PF, Judge ME, Andersen AS, Diers I, Egel-Mitani M, Hastrup S (1998) Purification and characterisation of a new hypothalamic satiety peptide, cocaine and amphetamine regulated transcript (CART), produced in yeast. *FEBS Lett* 428: 263–268
63 Asnicar MA, Smith DP, Yang DD, Heiman ML, Fox N, Chen YF, Hsiung HM, Koster A (2001) Absence of cocaine- and amphetamine-regulated transcript results in obesity in mice fed a high caloric diet. *Endocrinol* 142: 4394–4400
64 Abbott CR, Rossi M, Wren AM, Murphy KG, Kennedy AR, Stanley SA, Zollner AN, Morgan DG, Morgan I, Ghatei MA et al. (2001) Evidence of an orexigenic role for cocaine- and amphetamine-regulated transcript after administration into discrete hypothalamic nuclei. *Endocrinology* 142: 3457–3463
65 Breivogel CS, Childers SR (1998) The functional neuroanatomy of brain cannabinoid receptors. *Neurobiological Disorders* 5: 417–431
66 Glass M, Dragunow M, Faull RLM (1997) Cannabinoid receptors in the human brain: a detailed anatomical and quantitative autoradiographic study in the fetal, neonatal and adult human brain. *Neuroscience* 77: 299–318
67 Harrold JA, Elliott JC, King PJ, Widdowson PS, Williams G (2002) Down-regulation of cannabinoid–1 (CB-1) receptors in specific extrahypothalamic regions of rats with dietary obesity: a role for endogenous cannabinoids in driving appetite for palatable food? *Brain Res* 952: 232–238
68 Devane WA, Hanus L, Breuer A, Pertwee RG, Stevenson LA, Griffin G, Gibson D, Mandelbaum A, Etinger A, Mechoulam R (1992) Isolation and structure of a brain constituent that binds to the cannabinoid receptor. *Science* 258: 1946–1949
69 Di Marzo V, Melck D, Bisogno T, De Petrocellis L (1998) Endocannabinoids: endogenous cannabinoid receptor ligands with neuromodulatory action. *Trends in Neuroscience* 21: 521–528
70 Stella N, Schweitzer P, Piomelli D (1997) A second endogenous cannabinoid that modulates long-term potentiation. *Nature* 388: 773–778
71 Hanus L, Abu-Lafi S, Fride E, Breuer A, Vogel Z, Shaley DE, Kustanovich I, Mechoulam R (2001) 2-arachidonyl glyceryl ether, an endogenous agonist of the cannabinoid CB1 receptor. *Proc Natl Acad Sci USA* 98: 3662–3665
72 Porter AC, Sauer JM, Knierman MD, Becker GW, Berna MJ, Bao J, Nomikos GG, Carter P, Bymaster FP, Leese AB et al. (2002) Characterization of a novel endocannabinoid, virodhamine, with antagonist activity at the CB1 receptor. *J Pharmacol Exp Ther* 301: 1020–1024
73 Ravinet Trillou C, Delgorge C, Menet C, Arnone M, Soubrie P (2004) CB1 cannabinoid receptor knockout in mice leads to leanness, resistance to diet-induced obesity and enhanced leptin sensitivity. *Int J Obes Relat Metab Disord* 28: 640–648
74 Williams CM, Rogers PJ, Kirkham TC (1998) Hyperphagia in pre-fed rats following oral delta 9-THC. *Physiol Behav* 15: 343–346
75 Williams CM, Kirkham TC (1999) Anandamide induces over-eating: mediation by central cannabinoid (CB1) receptors. *Psychopharmacology* 143: 315–317
76 Hao S, Ayraham Y, Mechoulam R, Berry EM (2000) Low dose anandamide affects food intake,

cognitive function, neurotransmitter and corticosterone levels in diet-restricted mice. *Eur J Pharmacol* 392: 147–156

77 Corp ES, Melville LD, Greenberg D, Gibbs J, Smith GP (1990) Effect of fourth ventricular neuropeptide Y and peptide YY on ingestive and other behaviors. *Am J Physiol* 25 9: R317–R323

78 Erickson JC, Clegg KE, Palmiter RD (1996) Sensitivity to leptin and susceptibility to seizures of mice lacking neuropeptide Y. *Nature* 381: 415–421

79 Colombo G, Agabio R, Diaz G, Lobina C, Reali R, Gessa GL (1998) Appetite suppression and weight loss after the cannabinoid antagonist SR 141716. *Life Sci* 63: 113–117

80 Rowland NE, Mukherjee M, Robertson K (2001) Effects of the cannabinoid receptor antagonist SR 141716, alone and in combination with dexfenfluramine or naloxone, on food intake in rats. *Psychopharmacology* 159: 111–116

81 Kirkham TC, Williams CM, Fezza D, Di Marzo V (2002) Endocannabinoid levels in rat limbic forebrain and hypothalamus in relation to fasting, feeding and satiation: stimulation of eating by 2-arachidonoyl glycerol. *Br J Pharmacol* 136: 550–557

82 Di Marzo V, Goparaju SK, Wang L, Batkai S, Jarai Z, Fezza F, Miura GI, Palmiter RD, Sugiura T, Kunos G (2001) Leptin-regulated endocannainoids are involved in maintaining food intake. *Nature* 410: 822–825

83 Wenger T, Jamali KA, Juaneda C, Leonardelli J, Tramu G (1997) Arachidonylethanolamide (anandamide) activates the parvocellular part of the hypothalamic paraventricular nucleus. *Biochem Biophys Res Commun* 237: 724–728

84 Patel NA, Moldow RL, Patel JA, Wu G, Chang SL (1998) Arachidonylethanolamide (AEA) activation of FOS proto-oncogene protein immunoreactivity in the rat brain. *Brain Res* 797: 225–233

85 Jamshidi N, Taylor DA (2001) Anandamide administration into the ventromedial hypothalamus stimulates appetite in rats. *Br J Pharmacol* 134: 1151–1154

86 Cota D, Marsicano G, Tschop M, Grubler Y, Flachskamm C, Schubert M, Auer D, Yassouridis A, Thone-Reineke C, Ortmann S et al. (2003) The endogenous cannabinoid system affects energy balance via central orexigenic drive and peripheral lipogenesis. *J Clin Invest* 112: 423–431

87 Verty AN, McFarlane JR, McGregor IS, Mallett PE (2004) Evidence for an interaction between CB1 cannabinoid and melanocortin MC4-R receptors in regulating food intake. *Endocrinology* 45: 3224–3231

88 Finkelstein DI, Reeves AK, Horne MK (1996) An electron microscopic tracer study of projections from the entopeduncular nucleus to the ventrolateral nucleus of the rat. *Neurosci Lett* 211: 33–36

89 Gorbachevskaia AL (1999) Projections from the substantia nigra, ventral tegmental area and amygdale to the palladium in dog brain. *Morfologiia* 115: 11–14

90 Pecina S, Berridge KC (2000) Opioid sites in nucleus accumbens shell mediate eating and hedonic 'liking' for food: map based on microinjection fos plumes. *Brain Res* 863: 71–86

91 Arnone M, Maruani J, Chaperon F, Thiebot MH, Poncelet M, Soubrie P, Le Fur G (1997) Selective inhibition of sucrose and ethanol intake by SR 141716, an antagonist of central cannabinoid (CB1) receptors. *Psychopharmacology* 132: 104–106

92 Simiand J, Keane M, Keane PE, Soubrie P (1998) SR 141716, a CB1 cannabinoid receptor antagonist, selectively reduces sweet food intake in marmosets. *Behav Pharmacol* 9: 179–181

93 Welch SP, Eads M (1999) Synergistic interactions of endogenous opioids and cannabinoid systems *Brain Res* 848: 183–190

94 Kirkham TC, Williams CM (2001) Synergistic effects of opioid and cannabinoid antagonists on food intake. *Psychopharmacology* 153: 267–270

95 Iverson LL (2000) *The science of marijuana*. Oxford University Press, Oxford

96 Vickers SP, Kennett GA (2005) Cannabinoids and the regulation of ingestive behaviour. *Curr Drug Targets* 6: 215–223

97 Van Gaal LF, Rissanen AM, Scheen AJ, Ziegler O, Rossner S; RIO-Europe Study Group (2005) Effects of the cannabinoid-1 receptor blocker rimonabant on weight reduction and cardiovascular risk factors in overweight patients: 1-year experience from the RIO-Europe study. *Lancet* 365: 1389–1397

98 Pi-Sunyer FX, Aronne LJ, Heshmati HM, Devin J, Rosenstock J; RIO-North America Study Group (2006) Effect of rimonabant, a cannabinoid-1 receptor blocker, on weight and cardiometabolic risk factors in overweight or obese patients: RIO-North America: a randomized controlled trial. *JAMA* 295 : 761–775

99 Despres JP, Golay A, Sjostrom L; Rimonabant in Obesity-Lipids Study Group (2005) Effects of rimonabant on metabolic risk factors in overweight patients with dyslipidemia. *N Engl J Med* 353:

2121–2134
100 Scheen AJ, Finer N, Hollander P, Jensen MD, Van Gaal LF; RIO-Diabetes Study Group (2006) Efficacy and tolerability of rimonabant in overweight or obese patients with type 2 diabetes: a randomised controlled study. *Lancet* 368: 1660–1672
101 Leibowitz SF, Shor-Posner G (1986) Brain serotonin and eating behavior. *Appetite* 7: 1–14
102 Finn PD, Cunningham MJ, Rickard DG, Clifton DK, Steiner RA (2001) Serotonergic neurons are targets for leptin in the monkey. *J Clin Endocrinol Metab* 86: 422–426
103 Tecott LH, Abdallah L (2003) Mouse genetic approaches to feeding regulation: serotonin 5-HT2C receptor mutant mice. *CNS Spectr* 8: 584–588
104 Woolley ML, Marsden CA, Fone KC (2004) 5-HT6 receptors. *Curr Drug Targets CNS Neurol Disord* 3: 59–79
105 Vickers SP, Dourish CT (2004) Serotonin receptor ligands and the treatment of obesity. *Curr Opin Investig Drugs* 5: 377–388
106 Shacham S, Marantz Y, Senderowitz H (2005) Novel 5-HT6 receptor antagonists for the treatment of obesity. *Obes Res* 13: A192
107 Niijima A (1981) Visceral afferents and metabolic function. *Diabetologia* 20: 325–330
108 Moinat M, Deng C, Muzzin P, Assimacopoulos-Jeannet F, Seydoux J, Dulloo AG, Giacobino JP (1995) Modulation of obesegene-expression in rat brown and white adipose tissues. *FEBS Lett* 373: 131–134
109 Bi S, Gavrilova O, Gong DW, Mason MM, Reitman M (1997) Identification of a placental enhancer for the human leptin gene. *J Biol Chem* 272: 30583–30588
110 Bado A, Levasseur S, Attoub S, Kermorgant S, Laigneau JP, Bortoluzzi MN, Moizo L, Lehy T, Guerre-Millo M, Le Marchand-Brustel Y et al. (1998) The stomach is a source of leptin. *Nature* 394: 790–793
111 Isse N, Ogawa Y, Tamura N, Masuzaki H, Mori K, Okazaki T, Satoh N, Shigemoto M, Yoshimasa Y, Nishi S et al. (1995) Structural organization and chromosomal assignment of the human obese gene. *J Biol Chem* 270: 27728–27733
112 Cinti S, Frederich RC, Zingaretti MC, De Matteis R, Flier JS, Lowell BB (1997) Immunohistochemical localization of leptin and uncoupling protein in white and brown adipose tissue. *Endocrinology* 138: 797–804
113 Masuzaki H, Ogawa Y, Hosoda K, Kawada T, Fushiki T, Nakao K (1995) Augmented expression of the obese gene in the adipose-tissue from rats fed high-fat diet. *Biochem Biophys Res Commun* 216: 355–358
114 Chua SC Jr, Chung WK, Wu-Peng XS, Zhang Y, Liu SM, Tartaglia L, Leibel RL (1996) Phenotypes of mouse diabetes and rat fatty due to mutations in the OB (leptin) receptor. *Science* 271: 994–996
115 Campfield LA, Smith FJ, Guisez Y, Devos R, Burn P (1995) Recombinant mouse *ob* protein-evidence for a peripheral signal linking adiposity and central neural networks. *Science* 269: 546–549
116 Romsos DR, Swick AG, Chrunyk BA, Cunningham D, Mistry AM (1996) Intracerebroventricular recombinant leptin decreases food-intake and increases metabolic-rate in ob/ob mice. *Faseb J* 10: 1287
117 Wang Q, Bing C, Al-Barazanji K, Mossakowaska DE, Wang XM, McBay DL, Neville WA, Taddayon M, Pickavance L, Dryden S et al. (1997) Interactions between leptin and hypothalamic neuropeptide Y neurons in the control of food intake and energy homeostasis in the rat. *Diabetes* 46: 335–341
118 Bjorbaek C, Uotani S, da Silva B, Flier JS (1997) Divergent signaling capacities of the long and short isoforms of the leptin receptor. *J Biol Chem* 272: 32686–32695
119 Banks WA, Kastin AJ, Huang W, Jaspan JB, Maness LM (1996) Leptin enters the brain by a saturable system independent of insulin. *Peptides* 17: 305–311
120 Mercer JG, Moar KM, Rayner VD, Trayhurn P, Hoggard N (1997) Regulation of leptin receptor and NPY gene expression in hypothalamus of leptin-treated obese (*ob/ob*) and cold-exposed lean mice. *FEBS Lett* 402: 185–188
121 Baskin D, Breininger J, Schwartz M (1999) Leptin receptor mRNA identifies a subpopulation of neuropeptide Y neurons activated by fasting in rat hypothalamus. *Diabetes* 48: 828–833
122 Caro JF, Kolaczynski JW, Nyce MR, Ohannesian JP, Opentanova I, Goldman WH, Lynn RB, Zhang PL, Sinha MK, Considine RV (1996) Decreased cerebrospinal-fluid/serum leptin ratio in obesity: a possible mechanism for leptin resistance. *Lancet* 348: 159–161
123 Heymsfield SB, Greenberg AS, Fujioka K, Dixon RM, Kushner R, Hunt T, Lubina JA, Patane J,

Self B, Hunt P et al. (1999) Recombinant leptin for weight loss in obese and lean adults: a randomized, controlled, dose-escalation trial. *JAMA* 282: 1568–1575
124 Eckel LA, Langhans W, Kahler A, Campfield LA, Smith FJ, Geary N (1998) Chronic administration of OB protein decreases food intake by selectively reducing meal size in female rats. *Am J Physiol* 275: R186–R193
125 Kahler A, Geary N, Eckel LA, Campfield LA, Smith FJ, Langhans W (1998) Chronic administration of OB protein decreases food intake by selectively reducing meal size in male rats. *Am J Physiol* 275: R180–R185
126 Hukshorn CJ, Saris WH, Westerterp-Plantenga MS, Farid AR, Smith FJ, Campfield LA (2000) Weekly subcutaneous pegylated recombinant native human leptin (PEG-OB) administration in obese men. *J Clin Endocrinol Metab* 85: 4003–4009
127 Stahl N, Farruggella TJ, Boulton TG, Zhong Z, Darnell JE Jr, Yancopoulos GD (1995) Choice of STATs and other substrates specified by modular tyrosine-based motifs in cytokine receptors. *Science* 267: 1349–1353
128 Xu B, Kalra PS, Moldawer LL, Kalra SP (1998) Increased appetite augments hypothalamic NPY Y1 receptor gene expression: effects of anorexigenic ciliary neurotropic factor. *Regul Pept* 75–76: 391–395
129 Gloaguen I, Costa P, Demartis A, Lazzaro D, Di Marco A, Graziani R, Paonessa G, Chen F, Rosenblum CI, Van der Ploeg LH et al. (1997) Ciliary neurotrophic factor corrects obesity and diabetes associated with leptin deficiency and resistance. *Proc Natl Acad Sci USA* 94: 6456–6461
130 Lambert PD, Anderson KD, Sleeman MW, Wong V, Tan J, Hijarunguru A, Corcoran TL, Murray JD, Thabet KE, Yancopoulos GD et al. (2001) Ciliary neurotrophic factor activates leptin-like pathways and reduces body fat, without cachexia or rebound weight gain, even in leptin-resistant obesity. *Proc Natl Acad Sci USA* 98: 4652–4657
131 Marsh DJ, Hollopeter G, Huszar D, Laufer R, Yagaloff KA, Fisher SL, Burn P, Palmiter RD (1999) Response of melanocortin-4 deficient mice to anorectic and orexigenic peptides. *Nat Genet* 21: 119–122
132 Kalra SP (2001) Circumventing leptin resistance for weight control. *Proc Natl Acad Sci USA* 98: 4279–4281
133 Ettinger MP, Littlejohn TW, Schwartz SL, Weiss SR, McIlwain HH, Heymsfield SB, Bray GA, Roberts WG, Heyman ER, Stambler N et al. (2003) Recombinant variant of ciliary neurotrophic factor for weight loss in obese adults: a randomized, dose-ranging study. *JAMA* 289: 1826–1832
134 Grasso P, Leinung MC, Ingher SP, Lee DW (1997) *In vivo* effects of leptin-related synthetic peptides on body weight and food intake in female ob/ob mice: localization of leptin activity to domains between amino acid residues 106–140. *Endocrinology* 138: 1413–1418
135 Grasso P, Leinung MC, Lee DW (1999) Epitope mapping of secreted mouse leptin utilizing peripherally administered synthetic peptides. *Regul Pept* 85: 93–100
136 Lee DW, Leinung MC, Rozhavskaya-Arena M, Grasso P (2002) Leptin and the treatment of obesity: its current status. *Eur J Pharmacol* 440: 129–139
137 Grasso P, White DW, Tartaglia LA, Leinung MC, Lee DW (1999) Inhibitory effects of leptin-related synthetic peptide 116–130 on food intake and body weight gain in female C57BL/6J ob/ob mice may not be mediated by peptide activation of the long isoform of the leptin receptor. *Diabetes* 48: 2204–2209
138 Durrance A (2003) Obesity and related disorders – SMi Conference. Fat is no longer a feminist issue. *IDrugs* 6: 222–223
139 Brief DJ, Davis JD (1984) Reduction of food intake and body weight by chronic intraventricular insulin infusion. *Brain Res Bull* 12: 571–575
140 Menendez JA, Atrens DM (1991) Insulin and the paraventricular hypothalamus: modulation of energy balance. *Brain Res* 555: 193–201
141 Strubbe JH, Mein CG (1977) Increased feeding in response to bilateral injection of insulin antibodies in the VMH. *Physiol Behav* 19: 309–313
142 Schwartz MW, Marks JL, Sipols AJ, Woods SC, Kahn SE, Porte D Jr, (1991) Central insulin administration reduces neuropeptide Y mRNA expression in the arcuate nucleus of food-deprived lean (Fa/Fa) but not obese (fa/fa) Zucker rats. *Endocrinology* 128: 2645–2647
143 Sipols AJ, Baskin DG, Schwartz MW (1995) Effect of intracerebroventricular insulin infusion on diabetic hyperphagia and hypothalamic neuropeptide gene expression. *Diabetes* 44: 147–151
144 Schwartz MW, Sipols AJ, Marks JL, Sanacora G, White JD, Scheurink A, Kahn SE, Baskin DG, Woods SC, Figlewicz DP et al. (1992) Inhibition of hypothalamic neuropeptide Y gene expres-

sion by insulin. *Endocrinology* 130: 3608–3616
145 Bruning JC, Gautam D, Burks DJ, Gillette J, Schubert M, Orban PC, Klein R, Krone W, Muller-Wieland D, Kahn CR (2000) Role of brain insulin receptor in control of body weight and reproduction. *Science* 289: 2122–2125
146 Air EL, Strowski MZ, Benoit SC, Conarello SL, Salituro GM, Guan XM, Liu K, Woods SC, Zhang BB (2002) Small molecule insulin mimetics reduce food intake and body weight and prevent development of obesity. *Nat Med* 8: 303
147 Moran TH, Schwartz GJ (1994) Neurobiology of cholecystokinin. *Crit Rev Neurobiol* 9: 1–28
148 West DB, Fey D, Woods SC (1984) Cholecystokinin persistently suppresses meal size but not food intake in free-feeding rats. *Am J Physiol* 246: R776–R787
149 Matson CA, Ritter RC (1999) Long-term CCK-leptin synergy suggests a role for CCK in the regulation of body weight. *Am J Physiol* 276: R1038–R1045
150 Figlewicz DP, Stein LJ, West D, Porte D Jr, Woods SC (1986) Intracisternal insulin alters sensitivity to CCK-induced meal suppression in baboons. *Am J Physiol* 250: R856–R860
151 Konturek PC, Konturek SJ, Brzozowski T, Hahn EG (1999) Gastroprotection and control of food intake by leptin. Comparison with cholecystokinin and prostaglandins. *J Physiol Pharmacol* 50: 39–48
152 Brzozowski T, Konturek PC, Konturek SJ, Pajdo R, Duda A, Pierzchalski P, Bielanski W, Hahn EG (1999) Leptin in gastroprotection induced by cholecystokinin or by a meal. Role of vagal and sensory nerves and nitric oxide. *Eur J Pharmacol* 374: 263–276
153 Fan W, Ellacott KL, Halatchev IG, Takahashi K, Yu P, Cone RD (2004) Cholecystokinin-mediated suppression of feeding involves the brainstem melanocortin system. *Nat Neurosci* 7: 335–336
154 Dourish CT, Rycroft W, Iversen SD (1989) Postponement of satiety by blockade of brain cholecystokinin (CCK-B) receptors. *Science* 245: 1509–1511
155 Moran TH, Latz LF, Plata-Salaman CR, Schwartz GJ (1998) Disordered food intake and obesity in rats lacking cholecystokinin A receptors. *Am J Physiol* 274: R618–R625
156 Tschop M, Smiley DL, Heiman ML (2000) Ghrelin induces adiposity in rodents. *Nature* 407: 908–913
157 Nakazato M, Murakami N, Date Y, Kojima M, Matsuo H, Kangawa K, Matsukura S (2001) A role for ghrelin in the central regulation of feeding. *Nature* 409: 194–198
158 Wren AM, Small CJ, Abbott CR, Dhillo WS, Seal LJ, Cohen MA, Batterham RL, Taheri S, Stanley SA, Ghatei MA et al. (2001) Ghrelin causes hyperphagia and obesity in rats. *Diabetes* 50: 2540–2547
159 Shintani M, Ogawa Y, Ebihara K, Aizawa-Abe M, Miyanaga F, Takaya K, Hayashi T, Inoue G, Hosoda K, Kojima M et al. (2001) Ghrelin, an endogenous growth hormone secretagogue, is a novel orexigenic peptide that antagonizes leptin action through the activation of hypothalamic neuropeptide Y/Y1 receptor pathway. *Diabetes* 50: 227–232
160 Wren AM, Seal LJ, Cohen MA, Brynes AE, Frost GS, Murphy KG, Dhillo WS, Ghatei MA, Bloom SR (2001) Ghrelin enhances appetite and increases food intake in humans. *J Clin Endocrinol Metab* 86: 5992
161 Cummings DE, Purnell JQ, Frayo RS, Schmidova K, Wisse BE, Weigle DS (2001) A preprandial rise in plasma ghrelin levels suggests a role in meal initiation in humans. *Diabetes* 50: 1714–1719
162 Ariyasu H, Takaya K, Tagami T, Ogawa Y, Hosoda K, Akamizu T, Suda M, Koh T, Natsui K, Toyooka S et al. (2001) Stomach is a major source of circulating ghrelin, and feeding state determines plasma ghrelin-like immunoreactivity levels in humans. *J Clin Endocrinol Metab* 86: 4753–4758
163 Pinkney J, Williams G (2002) Ghrelin gets hungry. *Lancet* 359: 1360–1361
164 Tschop M, Smiley DL, Heiman ML (2000) Ghrelin induces adiposity in rodents. *Nature* 407: 908–913
165 English PJ, Ghatei MA, Malik IA, Bloom SR, Wilding JP (2002) Food fails to suppress ghrelin levels in obese humans. *J Clin Endocrinol Metab* 87: 2984
166 DelParigi A, Tschop M, Heiman ML, Salbe AD, Vozarova B, Sell SM, Bunt JC, Tataranni PA (2002) High circulating ghrelin: a potential cause for hyperphagia and obesity in prader-willi syndrome. *J Clin Endocrinol Metab* 87: 5461–5464
167 Asakawa A, Inui A, Kaga T, Katsuura G, Fujimiya M, Fujino MA, Kasuga M (2003) Antagonism of ghrelin receptor reduces food intake and body weight gain in mice. *Gut* 52: 947–952
168 Sesmilo G, Biller BM, Llevadot J, Hayden D, Hanson G, Rifai N, Klibanski A (2000) Effects of

growth hormone administration on inflammatory and other cardiovascular risk markers in men with growth hormone deficiency. A randomized, controlled clinical trial. *Ann Intern Med* 133: 111–122

169 Helmling S, Maasch C, Eulberg D, Buchner K, Schroder W, Lange C, Vonhoff S, Wlotzka B, Tschop MH, Rosewicz S et al. (2004) Inhibition of ghrelin action *in vitro* and *in vivo* by an RNA-Spiegelmer. *Proc Natl Acad Sci USA* 101: 13174–13179

170 Howard AD, Feighner SD, Cully DF, Arena JP, Liberator PA, Rosenblum CI, Hamelin M, Hreniuk DL, Palyha OC, Anderson J et al. (1996) A receptor in pituitary and hypothalamus that functions in growth hormone release. *Science* 273: 974–977

171 Smith RG, Pong SS, Hickey G, Jacks T, Cheng K, Leonard R, Cohen CJ, Arena JP, Chang CH, Drisko J et al. (1996) Modulation of pulsatile GH release through a novel receptor in hypothalamus and pituitary gland. *Recent Prog Horm Res* 51: 261–285

172 Pedersen-Bjergaard U, Host U, Kelbaek H, Schifter S, Rehfeld JF, Faber J, Christensen NJ (1996) Influence of meal composition on postprandial peripheral plasma concentrations of vasoactive peptides in man. *Scand J Clin Lab Invest* 56: 497–503

173 Halatchev IG, Ellacott KL, Fan W, Cone RD (2004) Peptide YY3–36 inhibits food intake in mice through a melanocortin-4 receptor-independent mechanism. *Endocrinology* 145: 2585–2590

174 Batterham RL, Cowley MA, Small CJ, Herzog H, Cohen MA, Dakin CL, Wren AM, Brynes AE, Low MJ, Ghatei MA et al. (2002) Gut hormone PYY(3–36) physiologically inhibits food intake. *Nature* 418: 650–654

175 Batterham RL, Cohen MA, Ellis SM, Le Roux CW, Withers DJ, Frost GS, Ghatei MA, Bloom SR (2003) Inhibition of food intake in obese subjects by peptide YY3–36. *N Engl J Med* 349: 941–948

Pharmacotherapy of Obesity
Edited by J.P.H. Wilding

Intestinal lipase inhibitors

John P.H. Wilding

Diabetes and Endocrinology Clinical Research Group, Clinical Sciences Centre, University Hospital Aintree, Longmoor Lane, Liverpool L9 7AL, UK

Introduction

Given that malabsorptive surgery is such an effective treatment for obesity, it is perhaps not surprising that attempts have been made to use drugs that inhibit the absorption of macronutrients to treat obesity. Acarbose, an inhibitor of carbohydrate absorption has been shown to have modest efficacy in the treatment of diabetes, but does not cause weight loss. Dietary fat is the most energy dense macronutrient, and blockade of absorption of fat is therefore a logical target for an anti-obesity drug. Orlistat, an inhibitor of pancreatic and intestinal lipases, was originally developed as a possible treatment for dyslipidaemia, but the focus switched to obesity once its ability to promote a negative energy balance was recognised.

Pharmacology

Orlistat is a white to off-white crystalline powder which is a chemically synthesised derivative of lipstatin, a natural product of the fungus *Streptomyces toxytricini* [1]. It contains a N-formyl-L-leucine ester side chain and a beta-lactone ring structure incorporated into a hydrocarbon backbone. The beta-lactone ring is essential for its lipase-inhibiting activity [2].

Orlistat is a potent inhibitor of pancreatic and intestinal lipases; it is not appreciably absorbed systemically and results in inhibition of the absorption of approximately 30% of dietary triglyceride. It binds covalently as an ester to the serine molecule at position 152 of pancreatic lipase [3] which is close to the putative active site of this enzyme, in a ratio of 1:1. It has also been shown to inhibit other intestinal lipases, such as gastric lipase, carboxyl ester lipase (pancreatic cholesterol esterase), with inhibition of the latter being reversible, due to hydrolysis of the drug forming an inactive product [4]. Orlistat does not have appreciable activity against other digestive enzymes such as proteases and amylase, nor does it affect systemic lipases when given in therapeutic doses [5].

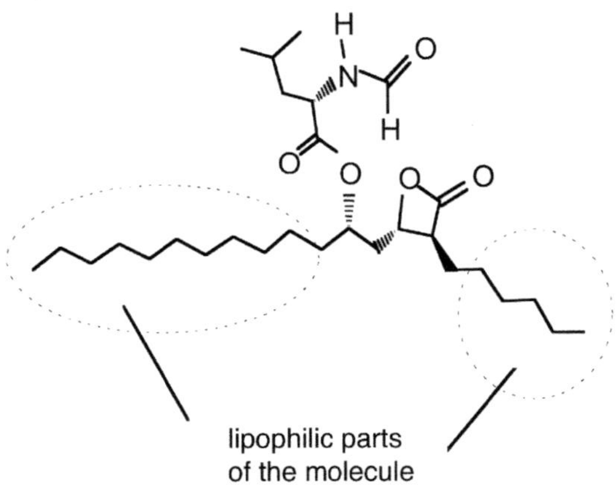

Figure 1. Structure of Orlistat. Orlistat (tetrahydrolipstatin; Xenical); $C_{29}H_{53}NO_5$; (S)-2-formylamino-4-methyl-pentanoic acid (S)-1-[[(2S, 3S)-3-hexyl-4-oxo-2-exetanyl]methyl]-dodecyl ester.

There are two major metabolites, designated M1 (4-member lactone ring hydrolysed); and M3 (M1 with N-formyl leucine ring cleaved) probably produced in the gut wall, which account for about 42% of detectable labelled activity in plasma following oral administration of radiolabelled orlistat [6]. These are very weak lipase inhibitors (M1 is 1,000-fold and M3 is 2,500-fold less active than the parent compound), plasma concentrations are very low and they are not considered to be of pharmacological importance [7, 8].

Oral administration of the drug results in inhibition of the absorption of dietary triglyceride in a dose dependent manner. The maximal inhibition that can be achieved is of about 32% of fat absorption, with about 30% of fat being excreted in the faeces at therapeutic doses [9]. Inhibition of fat absorption also reduces absorption of fat soluble vitamins, such as vitamin A, D, E, K and beta-carotene [10, 11].

Toxicology

Toxicological tests have been carried out in mice, rats and rabbits. No evidence of carcinogenicity was observed using the Ames test, a mammalian forward mutation assay (V79/HPRT), and *in vitro* clastogenesis assay in peripheral human lymphocytes, an unscheduled DNA synthesis assay (UDS) in rat hepatocytes in culture or in an *in vivo* mouse micronucleus test. No evidence of carcinogenicity or mutagenicity have been observed in normal rats and mice in doses up to 1,000 mg/kg/day and 1,500 mg/kg/day respectively or in female mice and rats predisposed to breast cancer. No teratogenic or embyropathic effects have been observed in rats or rabbits at doses up to 800 mg/kg/day (equivalent to 123 and 241 times the daily human dose calculated on a body surface area basis [8]).

Clinical pharmacology

The main pharmacological effect in humans after an oral dose of orlistat is a reduction in absorption of dietary fat, measured as the percentage of fat in the diet which is excreted in the faeces. In untreated healthy subjects this is less than 5%; administration of orlistat increases faecal fat excretion in a dose dependent manner. This has been studied in normal subjects in doses ranging from 10 mg to 400 mg given orally three times daily with meals. The steepest part of the dose response curve is seen between 30 and 400 mg daily, and the maximal effect (inhibition of 32% of dietary fat absorption) is seen at a dose of 400 mg daily, with no increase at higher doses. The daily dose that produced 50% of the maximal effect is 98 mg/day [9]. There are no differences in pharmacology between lean and obese subjects. The two identified metabolites, M1 and M3 are essentially pharmacologically inactive. There is little systemic absorption and no measurable effect on systemic lipase activity.

Weight loss was investigated in a 12 week dose-ranging study with doses ranging from 10 to 120 mg tds, with only the doses of 60 mg and 120 mg tds resulting in significant weight loss. Higher doses (up to 240 mg tds) do not improve efficacy [12].

Pharmacokinetics

Pharmacokinetics has been assessed in obese patients using a single 360 mg dose of ^{14}C-labelled orlistat. The majority (96.4 ± 18.1%) of orally administered orlistat is excreted unchanged in faeces. Less than 1.2% of a radioactive dose is systemically absorbed and the maximal plasma concentrations of radioactivity reached after a 360 mg dose in healthy volunteers were 150 ± 51 ng eq/ml at 6.8 ± 1.5 h. Orlistat has been measured in plasma from over 5,000 trial samples using a method employing high-performance liquid chromatography coupled with ion-spray tandem mass spectrometry. The lower detection limit of this assay is 0.1 ng/ml of plasma, and the lower limit of quantitation 0.2 ng/ml [13]. Orlistat is only sporadically measurable in plasma (i.e., most samples < 0.2 ng/ml) when given at therapeutic doses in clinical trials for periods of up to 2 years, with the highest concentrations being < 10 ng/ml, indicating that the drug does not accumulate in plasma [7].

The *in vivo* volume of distribution cannot be calculated because of the very low systemic absorption of the drug. *In vitro*, 99% of the drug is bound to plasma proteins (albumin and lipoproteins). Kinetics have not been studied in special groups, such as in children, the elderly and in renal and hepatic impairment because of the low systemic absorption of the drug. The majority (98.4%) of orally administered orlistat is excreted unchanged in the faeces. Some metabolism is thought to occur in the gut wall, to produce the M1 and M3 metabolites.

Drug interactions

There is an increased risk of does-related adverse effects, particularly rhabdomyolysis, in patients taking pravastatin. This is due to an increase in pravastatin concentrations. Absorption of fat soluble vitamins and β-carotene may be impaired. No interactions have been observed with digoxin, phenytoin, oral contraceptives, glibenclamide, nifedipine, or alcohol [14–19]. Patients taking warfarin or other anticoagulants should have their INR monitored, although there is no evidence that INR is altered by treatment with orlistat at therapeutic doses [17]. There is no information on co-administration with fibrates, acarbose or anorectic drugs.

Therapeutic use

Orlistat is indicated as a treatment for obesity, defined as a body mass index (BMI, calculated as weight in kg/height in m^2) greater than or equal to 30 kg/m^2; orlistat may also be considered in overweight (BMI greater than or equal to 28 kg/m^2) with associated risk factors such as type 2 diabetes, hypertension, hyperlipidaemia. In many countries, including the UK, it is recommended that treatment should be discontinued if treatment has not resulted in 5% weight loss after 12 weeks of drug treatment. The main contraindications to orlistat are chronic malabsorption syndromes, cholestasis, breast-feeding and hypersensitivity to any component of the product.

It is given as capsules, taken orally in a dose of 120 mg three times daily, taken before, during or up to 1 h after each main meal. Patients should receive advice to keep to a diet which contains less that 30% of calories from fat, and ideally the fat content of the diet should be evenly divided between three main meals. If a meal contains no fat, the dose can be omitted. A meal with a very high fat content will increase the risk of gastrointestinal side effects.

As the drug may reduce levels of fat soluble vitamins (A, D, E, K) and β-carotene, patients are advised to have a diet which is rich in fruit and vegetables and use of a multivitamin supplement should be considered; this should be taken at least 2 h after the administration of orlistat.

Efficacy – obesity clinical trials

Several well-designed randomised clinical trials have demonstrated that treatment of obesity with orlistat is safe and well tolerated by most patients [20, 21]. The resultant mean weight loss over placebo is about 3.5 kg after 2 years treatment, but about twice as many patients achieve clinically beneficial weight loss of 5% with orlistat treatment (58% *versus* 32%); the proportion achieving 10% with treatment is also greater (39% *versus* 18%). Modest reductions in several cardiovascular risk factors have been found, including total and LDL cholesterol, fasting insulin and blood pressure. However patients with raised blood pressure, diabetes or significant hypercholesterolaemia were excluded from these studies.

A number of studies have been carried out, ranging in duration from a few months to 4 years in duration. The pivotal Phase III double-blind placebo-controlled clinical trials using orlistat to treat obesity have been reported in full; the first of these was in European patients and studied 743 patients [20]. After a dietary lead-in period of 4 weeks, patients were randomly allocated to treatment with either orlistat 120 mg tds or placebo. A diet calculated to provide 600 kcal/day deficit was prescribed for the first 12 months, a weight maintenance diet was recommended for the second 12 months, following a further randomisation to either placebo or orlistat. The primary outcome measure was weight loss, and secondary outcome measures included co-morbidities such as blood pressure, lipids, insulin and glucose, although patients with significant hypertension, hyperlipidaemia or diabetes were excluded from the study. Patients assigned orlistat lost an average of 10.3 kg over the first year *versus* 6.1 kg in the placebo group. Weight gain during the weight maintenance period (year 2) was less in those assigned to continue orlistat compared to those treated with placebo. At the end of the first year, 39% of orlistat-treated patients had lost at least 10% of body weight compared to 18% of placebo.

Total and LDL cholesterol, fasting glucose and insulin also improved with orlistat treatment compared to placebo; these effects were relatively modest (e.g., a 0.12 mmol fall in cholesterol at the end of year 2 and a 12.7% fall in the LDL/HDL ratio), but should be considered in the light of increases in levels of these risk factors in the placebo group. The effects seen with orlistat on lipids were greater than would be expected with weight loss alone, and may reflect the mode of action of the drug.

The results of a similar 2 year double-blind, placebo-controlled study carried out in the US have also been reported [22]. Over 900 patients were randomised to treatment; two-thirds were treated with orlistat for the first year, the remainder placebo. During the second year, half of the orlistat-treated patients were re-randomised to placebo, orlistat 60 mg tds, or orlistat 120 mg tds. All patients received dietary and lifestyle advice. The results were broadly similar to the European study, with a weight loss of $8.8 \pm 0.4\%$ *versus* $5.8 \pm 0.7\%$; $p < 0.001$ after 1 year of therapy. Moreover, a significantly greater proportion of orlistat-treated patients lost >5% or >10% of their initial body weight compared with placebo (65.7% *versus* 43.6%, $p < 0.01$ and 38.9% *versus* 24.8%, $p = 0.004$, respectively). During the second year, there was less weight regain in the patients treated with orlistat 120 mg tds (3.2 ± 0.45 kg *versus* 5.63 ± 0.42 kg), but 60 mg tds was relatively ineffective. Over one-third of patients (34.1%) who received orlistat 120 mg for 2 years maintained weight loss of >10%, twice as many as in the placebo group (17.5%; $p = 0.02$).

Modest improvements were seen with a number of cardiovascular risk factors, including total cholesterol (0.3 mmol/l fall compared to placebo), LDL-

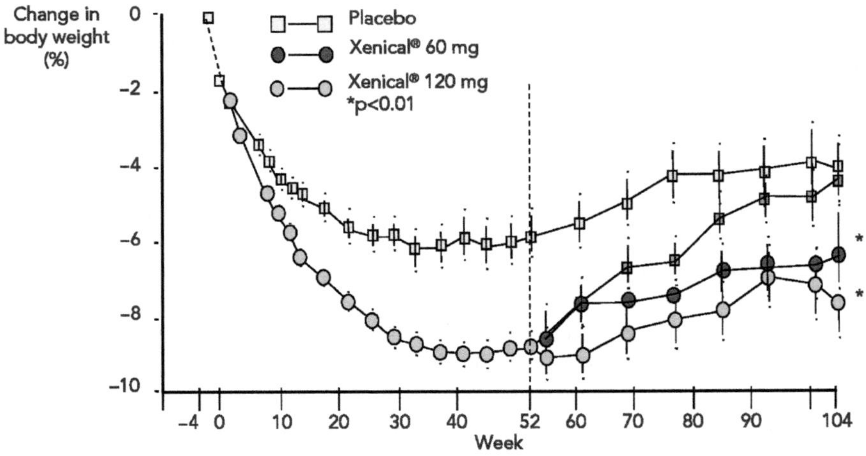

Figure 2. Weight loss and regain on stopping treatment.

cholesterol (0.15 mmol/l fall), fasting insulin (21% fall compared to no change with placebo), and blood pressure (approx 1 mmHg fall in systolic and diastolic blood pressure). Withdrawals due to GI side effects were 7% in the orlistat treated group.

The amount of weight loss achieved with orlistat is slightly less than reported with sibutramine or rimonabant [23–25], although differences in trial design may account for some of these differences, the placebo-subtracted weight loss is certainly greater with the other agents. A recent study suggested that responders to orlistat were more likely to have personality traits of personality dimension 'conscientiousness' (e.g., 'order' and 'deliberation'), which may reflect the fact that a certain amount of personal discipline is required to use orlistat in an effective way (avoiding high fat meals, and taking the drug regularly with each meal). Whether this observation is useful in clinical practice remains to be seen [26].

Efficacy in obesity with co-morbidity

A number of studies in type 2 diabetic patients have been reported and suggest that the absolute weight loss tends to be less than in non-diabetic subjects. One study however does suggest there may be a small, independent effect on insulin sensitivity [27]. Whether patients are on metformin or SU background therapy, the difference from placebo was similar to the non-diabetic subjects, and is associated with an approximate 0.4% reduction in HbA_{1c} [28]. For example, in a 1 year double-blind placebo-controlled study of orlistat treatment carried out in 391 patients with type 2 diabetes: The patients were all overweight or obese (BMI 28–40 kg/m^2) and had stable diabetes treated with sulphonylureas. Weight loss in these diabetic patients was less than in non-diabetic obese subjects (mean of 6.2 kg with orlistat *versus* 4.3 kg in the placebo group); nevertheless, this resulted in improvements in fasting glucose (fall of 0.02 mmol *versus* 0.54 mmol increase in the placebo group), HbA_{1c} (0.28% fall *versus* rise of 0.18% in placebo group), insulin and lipids. Dose requirement for sulphonylurea therapy was also reduced, and 11.7% of orlistat-treated patients were able to stop treatment with sulphonylureas.

Two 1-year multimorbidity studies have also been reported. These enrolled patients with one or more cardiovascular risk factors, such as hypertension and dyslipidaemia. Weight loss in these studies was consistent with the other Phase III data reported above, and was associated with improvements across a range of risk factors, including blood pressure, LDL cholesterol and triglycerides [29, 30]. A study in patients with hypertension showed reductions in blood pressure [31]. The 4 year XENDOS study investigated the effects of orlistat in 3,305 Swedish obese patients, enriched for those at risk of type 2 diabetes, particularly with impaired glucose tolerance, which was present in 20% of the study population [32]. Although there was some weight regain in both placebo and orlistat groups in the XENDOS study, the placebo-subtracted weight loss remained approximately constant and amounted to 2.8 kg (last observation carried forward analysis) after 4 years of treatment. Overall there was a 37% relative risk reduction in the development of type 2 diabetes; this benefit was confined to those patients who had impaired glucose tolerance at baseline. The absolute rates of progression to diabetes in this group were 28.8% in the placebo group, compared to 18.8% in the orlistat group (a relative risk reduction of 45%). It should be remembered that this effect is in addition to the effect of diet and lifestyle which has also been shown to reduce the risk of diabetes progression over 4 years in a number of diabetes prevention trials, such as the Finnish and American diabetes prevention studies. Improvements in other cardiovascular risk factors were also seen with orlistat in the XENDOS study, this includes improvements in blood pressure, total and LDL cholesterol, fasting glucose, insulin and PAI-1. However fibrinogen fell slightly less than with placebo, and the rise in HDL cholesterol was less, although the LDL:HDL ratio improved (Tab. 1). It is not known whether these changes are associated with a decreased risk of vascular disease or cardiovascular death as no endpoint studies with orlistat have been conducted.

Use in children and adolescents

Orlistat has also been investigated in the treatment of adolescents. Although it is more difficult to assess weight changes in growing children, the effects seen in clinical trials are broadly similar to those seen in adult populations [33]. The side-effect profile may limit enthusiasm for the use of orlistat in this population, although it has been approved for use in Europe for adolescents with obesity.

Table 1. Effects of orlistat on selected cardiovascular risk factors in the XENDOS study

Risk factor	Placebo	Orlistat
Diastolic BP	−1.9	−2.6
Systolic BP	−3.4	−4.9
Total cholesterol	−2.3	−7.9
LDL cholesterol	−5.1	−12.8
HDL cholesterol	+9.1	+6.5
Triglycerides	2.9	2.4
Fasting glucose	0.2	0.1
Fasting insulin	−20.6	−32
PAI-1	0.1	−3.0
Fibrinogen	−0.5	−0.4

All differences at least $p < 0.05$

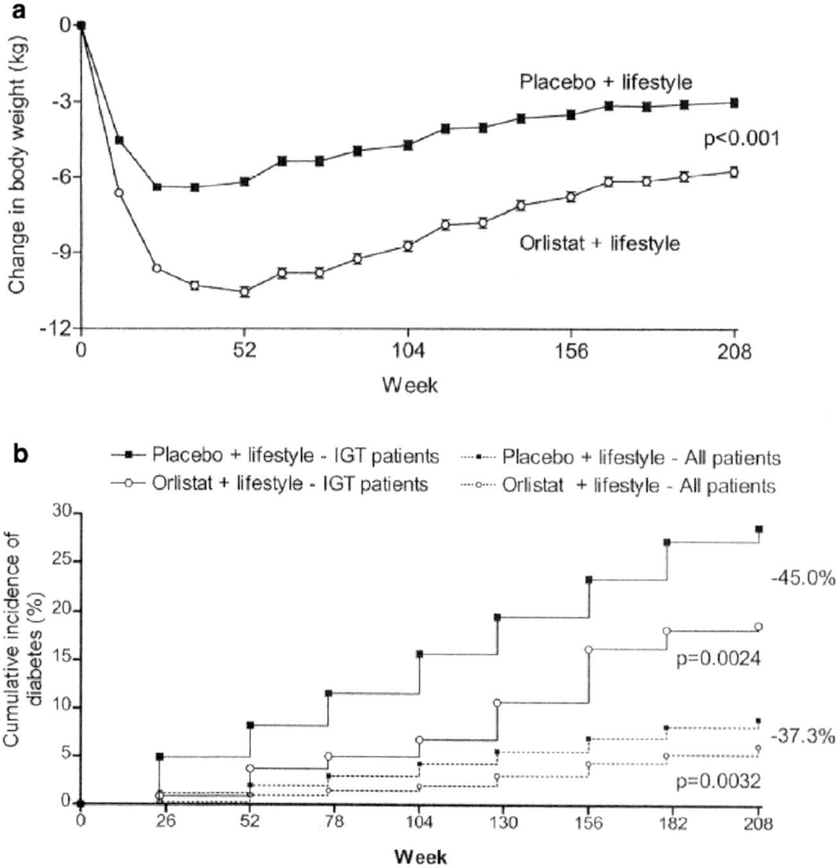

Figure 3. (a) Weight loss and (b) diabetes prevention in the XENDOS study [32].

Adverse reactions

Most of the adverse reactions are predictable from the mode of action of the drug, and are gastrointestinal in nature. It is also of note that these declined markedly during the second year of clinical trials with the drug.

GI adverse effects

The commonest adverse effects in this category were oily spotting (26.6% in first year; 4.4% in second year), flatus with discharge (23.9% and 2.1%), faecal urgency (22.1% and 2.8%), fatty or oily stool (20% and 5.5%). Oily evacuation and increased defaecation were reported by 11.9 and 10.8% of patients in the first year and 2.3 and 2.6% in the second year. 7.8% reported faecal incontinence, which fell to 1.8% in the second year.

Most patients only reported a single episode of such an event, and these only led to withdrawal from treatment in about 1% of cases. The incidence of such events appears to be related to the fat content of the drug; it is thought that patients soon learn to avoid high fat meals while on treatment. This may also improve efficacy, by encouraging patients to remain on a low fat diet.

Reduced plasma levels of fat-soluble vitamins and β-carotene

As the drug inhibits absorption of fat it also inhibits absorption of a proportion of fat soluble vitamins, specifically vitamins A, D, E, K and β-carotene. During clinical trials lasting up to 2 years, plasma levels of these substances fell slightly shortly after initiation of treatment, and then reached a plateau level which was maintained throughout the study. Less than 6% of patients had two values below the lower limit of normal for any one of these parameters during these studies. PTH levels and prothrombin time, measured as indices of vitamin D and K actions) did not change during treatment [20].

Cetilistat

A second lipase inhibitor, cetilistat is in clinical development. Limited clinical information is available for this agent; its efficacy appears similar to orlistat, but with a low incidence of GI side effects in a 12 week dose-ranging study, compared to what might be expected with orlistat [34]. The results of ongoing Phase III trials with cetilistat are awaited with interest.

References

1 Weibel EK, Hadvary P, Hochuli E (1987) Lipstatin, an inhbitor of pancreatic lipase, produced by *Streptomyces toxytricini*. I. Producing organism, fermentation, isolation and biological activity. *J Antibiot* 40: 1081–1085
2 Hochuli E, Kupfer E, Maurer R (1987) Lipstatin, an inhibitor of pancreatic lipase, produced by *Streptomyces toxitricini*. II Chemistry and Structure elucidation. *J Antibiot* 40: 1086–1091
3 Hadvary P, Sidler W, Meister W, Vetter W, Wolfer H (1991) The lipase inhibitor tetrahydrolipstatin binds covalently to the putative active-site serine of pancreatic lipase. *J Biol Chem* 266: 2021–2027
4 Borgstrom B (1988) Mode of action of tetrahydrolipstatin: A derivative of the naturally occurring lipase inhibitor lipstatin. *Biochem Biophys Acta* 962: 308–316
5 Guerciolini R (1997) Mode of action of orlistat. *Int J Obesity* 21: S12–S23
6 Zhi JG, Melia AT, Funk C, VigerChougnet A, Hopfgartner G, Lausecker B, Wang K, Fulton JS, Gabriel L, Mulligan TE (1996) Metabolic profiles of minimally absorbed orlistat in obese/overweight volunteers. *J Clin Pharmacol* 36: 1006–1011
7 Zhi JG, Melia AT, Eggers H, Joly R, Patel IH (1995) Review of limited systemic absorption of orlistat, a lipase inhibitor, in healthy-human volunteers. *J Clin Pharmacol* 35: 1103–1108
8 Orlistat Product Monograph 1 (1998) F-Hoffmann-La Roche Ltd.
9 Zhi J, Melia AT, Guerciolini R, Chung J, Kinberg J, Hauptman JB, Patel IH (1994) Retrospective population-based analysis of the dose-response (Fecal fat excretion) Relationship of orlistat in normal and obese volunteers. *Clin Pharmacol Therapeutics* 56: 82–85
10 Melia AT, KossTwardy SG, Zhi JG (1996) The effect of orlistat, an inhibitor of dietary fat absorption, on the absorption of vitamins A and E in healthy volunteers. *J Clin Pharmacol* 36: 647–653
11 Zhi JG, Melia AT, KossTwardy SG, Arora S, Patel IH (1996) The effect of orlistat, an inhibitor of dietary fat absorption, on the pharmacokinetics of beta-carotene in healthy volunteers. *J Clin Pharmacol* 36: 152–159
12 Drent ML, Larsson I, Williamolsson T, Quaade F, Czubayko F, Vonbergmann K, Strobel W, Sjostrom L, vanderVeen EA (1995) Orlistat (Ro-18-0647), A lipase inhibitor, in the treatment of human obesity – a multiple-dose study. *Int J Obesity* 19: 221–226
13 Bennett PK, Li YT, Edom R, Henion J (1997) Quantitative determination of Orlistat (tetrahydrolipostatin, Ro 18-0647) in human plasma by high-performance liquid chromatography coupled with ion spray tandem mass spectrometry. *J Mass Spectrometry* 32: 739–749
14 Weber C, Tam YK, SchmidtkeSchrezenmeier G, Jonkmann JHG, vanBrummelen P (1996) Effect of the lipase inhibitor orlistat on the pharmacokinetics of four different antihypertensive drugs in healthy volunteers. *Eur J Clin Pharmacol* 51: 87–90
15 Melia AT, Mulligan TE, Zhi JG (1996) The effect of orlistat on the pharmacokinetics of phenytoin in healthy volunteers. *J Clin Pharmacol* 36: 654–658
16 Melia AT, Zhi JG, KossTwardy SG, Min BH, Smith BL, Freundlich NL, Arora S, Passe SM (1995) The influence of reduced dietary-fat absorption induced by orlistat on the pharmacokinetics of digoxin in healthy-volunteers. *J Clin Pharmacol* 35: 840–843
17 Zhi J, Melia AT, KossTwardy SG, Min B, Guerciolini R, Freundlich NL, Milla G, Patel IH (1995) The influence of orlistat on the pharmacokinetics and pharmacodynamics of glyburide in healthy-volunteers. *J Clin Pharmacol* 35: 521–525
18 Hartmann D, Guzelban C, Zuiderwijk PBM, Odink J (1996) Lack of interaction between orlistat and oral contraceptives. *Eur J Clin Pharmacol* 50: 421–424
19 Melia AT, Zhi J, Zelasko R, Hartmann D, Guzelhan C, Guerciolini R, Odink J (1998) The interaction of the lipase inhibitor orlistat with ethanol in healthy volunteers. *Eur J Clin Pharmacol* 54: 773–777
20 Sjostrom L, Rissanen A, Andersen T, Boldrin M, Golay A, Koppeschaar HPF, Krempf M (1998) Randomised placebo-controlled trial of orlistat for weight loss and prevention of weight regain in obese patients. *Lancet* 352: 167–172
21 Davidson MH, Hauptman J, DiGirolamo M, Foreyt JP, Halsted CH, Heber D, Heimburger DC, Lucas CP, Robbins DC, Chung J et al. (1999) Weight control and risk factor reduction in obese subjects treated for 2 years with orlistat – A randomized controlled trial. *JAMA* 281: 235–242
22 Davidson M (1997) A 2 year, US, randomized, controlled study of orlistat, a gastrointestinal lipase inhibitor, for obesity treatment. *JAMA* 96: 4119
23 Bray GA, Ryan DH, Gordon D, Heidingsfelder S, Cerise F, Wilson K (1996) A double-blind ran-

domized placebo-controlled trial of sibutramine. *Obes Res* 4(3): 263–270
24 James WPT, Astrup A, Finer N, Hilsted J, Kopelman P, Rossner S, Saris WHM, Van Gaal LF (2000) Effect of sibutramine on weight maintenance after weight loss: a randomised trial. *Lancet* 356(9248): 2119–2125
25 Despres JP, Golay A, Sjostrom L (2005) Effects of rimonabant on metabolic risk factors in overweight patients with dyslipidemia. *New Eng J Med* 353(20): 2121–2134
26 Elfhag K, Finer N, Rossner S (2007) Who will lose weight on orlistat and sibutramine? Psychological correlates of success. *Diabetes Obes Metab* 26 (ePub ahead of print)
27 Tiikkainen M, Bergholm R, Rissanen A, Aro A, Salminen I, Tamminen M, Teramo K, Yki-Jarvinen H (2004) Effects of equal weight loss with orlistat and placebo on body fat and serum fatty acid composition and insulin resistance in obese women. *Am J Clin Nutrition* 79(1): 22–30
28 Hollander PA, Elbein SC, Hirsch IB, Kelley D, McGill J, Taylor T, Weiss SR, Crockett SE, Kaplan RA, Comstock J et al. (1998) Role of orlistat in the treatment of obese patients with type 2 diabetes – A 1-year randomized double-blind study. *Diabetes Care* 21: 1288–1294
29 Broom I, Wilding J, Stott P, Myers N (2002) Randomised trial of the effect of orlistat on body weight and cardiovascular disease risk profile in obese patients: UK multimorbidity study. *Int J Clin Practice* 56(7): 494–499
30 Lindgarde F (2000) The effect of orlistat on body weight and coronary heart disease risk profile in obese patients: The Swedish Multimorbidity Study. *J Internal Med* 248(3): 245–254
31 Sharma AM, Golay A (2001) Effect of weight management with orlistat on blood pressure in obese patients with hypertension. *Obesity Res* 9: 189S
32 Torgerson JS, Hauptman J, Boldrin MN, Sjostrom L (2004) XENical in the prevention of diabetes in obese subjects (XENDOS) study. *Diabetes Care* 27(1): 155–161
33 Chanoine JP, Hampl S, Jensen C, Boldrin M, Hauptman J (2005) Effect of orlistat on weight and body composition in obese adolescents – A randomized controlled trial. *JAMA* 293(23): 2873–2883
34 Kopelman P, Bryson A, Hickling R, Rissanen A, Rossner S, Toubro S, Valensi P (2007) Cetilistat (ATL-962), a novel lipase inhibitor: a 12-week randomized, placebo-controlled study of weight reduction in obese patients. *Int J Obesity* 31(3): 494–499

Pharmacotherapy of Obesity
Edited by J.P.H. Wilding

Sibutramine

John P.H. Wilding

School of Clinical Sciences, Clinical Sciences Centre, University Hospital Aintree, Longmoor Lane, Liverpool L9 7AL, UK

Introduction

Concerns over the development of conditions such as primary pulmonary hypertension and valvular heart disease in patients taking serotonin releasing agents such as dexfenfluramine and fenfluramine led to searches for the agents that worked via serotonin but did not produce the same cardiovascular side effects. Sibutramine is a serotonin and noradrenaline reuptake inhibitor, pharmacologically distinct from these other drugs which are considered serotonin releasing agents, which was originally developed as an antidepressant [1]. Early clinical studies showed sibutramine to be largely devoid of antidepressant properties, but resulted in weight loss [2]. This therefore led to the development of sibutramine as an anti-obesity agent. Sibutramine has now been shown to be effective at producing weight loss in simple obesity and obesity associated with a range of co-morbidities including hypertension, dyslipidaemia and type 2 diabetes, and this chapter will describe its preclinical and clinical pharmacology, efficacy in clinical trials and describe the progress of long-term outcome trials with sibutramine, such as the Sibutramine Cardiovascular OUtcomes Trial (SCOUT study).

Pharmacology

Sibutramine (1-(1-(4-chlorophenyl)cyclobutyl)-N,N,3-trimethylbutan-1-amine) (Fig. 1) is an orally administered centrally-acting antiobesity agent. The site of action of sibutramine and its active metabolites are predominantly in the central nervous system where it has been shown in a number of *in vitro* studies to inhibit both serotonin and nor-adrenaline reuptake [1, 3, 4]. These effects are attributable to its active metabolites, M1 and M2, which have a half life of approximately 14–19 h. The absorption of sibutramine is not affected by food. There is little difference in the elimination half life between elderly and young patients. Elimination has not been shown to be effected by renal dysfunction but its use is contraindicated in severe liver disease as the drug is eliminated via metabolism in the liver [5]. Preclinical studies have shown that sibutramine

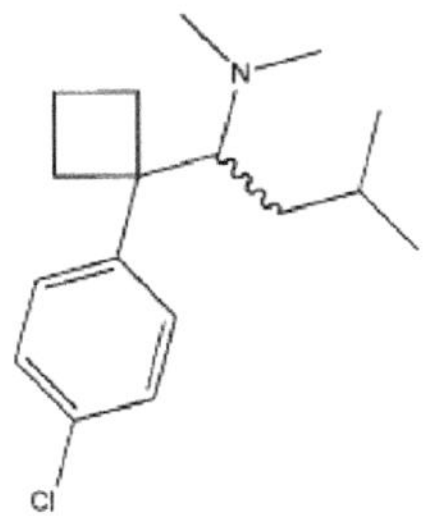

Figure 1. Structure of sibutramine (1-(1-(4-chlorophenyl)cyclobutyl)-N,N,3-trimethylbutan-1-amine).

is effective at reducing food intake in various animal models of obesity, including genetic models such as the ob/ob mouse, dietary induced obesity with palatable or via fat diet [6–9]. The effect on food intake in animal models has been shown to be due to enhancement of the natural satiety process rather than disruption of the satiety sequence as is seen with other drugs such as amphetamines and dexfenfluramine [8]. This effect on satiety has also been confirmed in human studies, which have also demonstrated a modest effect on body weight and respiratory quotient [10–12]. Sibutramine has much less effect on dopamine release compared to older anti-obesity drugs [13], nor does it bind to dopamine receptors which may partly explain its low potential for abuse in human studies [14]. The effects on metabolism have been shown in animals to be due to an increase in oxygen consumption which is likely mediated by increased sympathetic nervous system activity, via the β_3 adrenoreceptor [15, 16]. It is not clear whether the effect is relevant in humans, although one study has suggested that sibutramine prevents the usual decline in metabolic rate that is seen with weight loss, and other studies have shown that sibutramine alters respiratory quotient [11].

Clinical pharmacology

Sibutramine is well absorbed after a single oral dose, and undergoes extensive first pass metabolism to form its active amine metabolites, M2 and M1. Steady state concentrations are achieved within 4 days of administration of 15 mgs per day of sibutramine (the higher clinically used dose); there is no effect of renal impairment on sibutramine clearance, but c-max values were higher in patients with moderate hepatic impairment, and the drug is contra-indicated in those with severe hepatic impairment [5, 17].

Adverse effects

The most common adverse effects of sibutramine seen in clinical trials were headache, dry mouth, anorexia, insomnia and constipation. The side effect which has caused greatest concern to clinicians and regulators has been its effects on blood pressure and heart rate. Significant cardiovascular events such as significant hypertension, palpitations and tachycardia occur rarely in clinical trials, although these do occur at a higher rate than placebo. The numbers of patients withdrawn during clinical trials because of hypertension is under 1% and those because of tachycardia under 0.5%. The increase in heart rate with sibutramine treatment ranges from between 3 and 7 mmHg in clinical trials. The average rise in blood pressure is about 1–2 mmHg. Careful study of the cardiovascular effects of sibutramine have been carried out by Birkenfeld et al., who clearly demonstrated that although sibutramine has a peripheral effect to increase blood pressure, by increasing norepinephrine concentrations within the synapse, the central effect of sibutramine blockade of the norepinephrine transporters results in an attenuation of sympathetic outflow via activation of alpha 2 adrenoceptors [18] (Fig. 2). This clonidine-like sympatholyt-

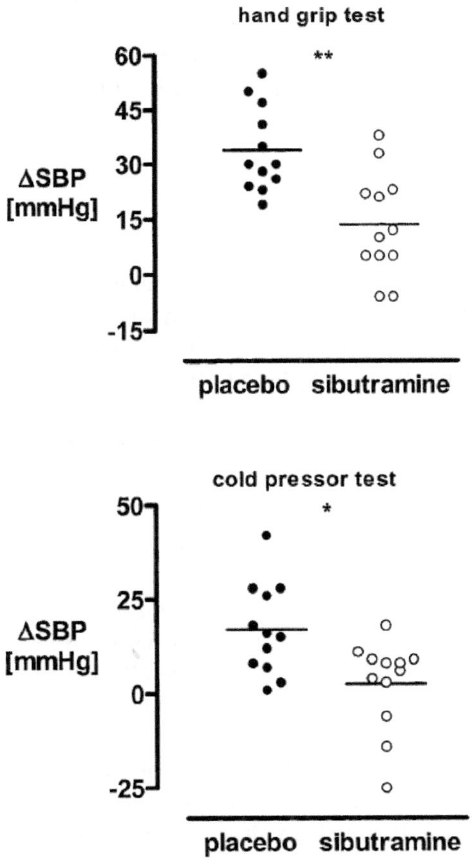

Figure 2. Paraoxical fall in blood pressure with sibutramine during hand grip and cold pressor tests indicating central clonidicine-like effect [18].

ic effect clearly modulates the peripheral pressor effect of sibutramine and can make it difficult to predict the effect of sibutramine on blood pressure in individual patients. Nevertheless, the rate of significant rises in blood pressure in sibutramine treated patients in clinical trials is low, even in a high risk population with established cardiovascular disease and/or diabetes, as was studied in the SCOUT study where less than 5% of patients experienced a rise of more than 10 mmHg in systolic or diastolic blood pressure or 10 bpm in heart rate during 6 weeks of open-label sibutramine treatment, and indeed blood pressure fell in the majority of patients [19, 20].

Clinical efficacy

In terms of its effects on food intake and body weight, sibutramine has been shown to produce dose-related weight loss which is distinguishable from placebo, at doses as low as 1 mg and has been shown to be effective up to 30 mg in an early dose ranging study [21]. As with other weight loss treatments, weight loss with sibutramine tends to plateau after about 6 months therapy, although this weight loss may continue for a little longer in patients with diabetes.

Therapeutic use

Sibutramine 10 and 15 mg tablets are indicated as an adjunctive therapy within a weight management programme for patients with obesity and a body mass index of greater than 30 kgs/m^2, or those with nutritional excess weight and a body mass index of greater than 27 kg/m^2 if other obesity-related risk factors such as type 2 diabetes or dyslipidaemia are present. It is important that patients are undertaking a weight management programme, and this should have failed to achieve weight loss of at least 5% over 3 months before starting sibutramine therapy. There is good evidence from clinical trials that greater weight loss is observed if drug therapy is given together with a behavioural support programme [22] (Fig. 3), and although the intensity of this programme has not been specified in detail, it is likely that weight loss will be greater in those patients who receive the greatest support. It is recommended that sibutramine treatment should only be continued if clinically significant weight loss is observed after 3 months of therapy, although some guidelines do suggest that this period of time may be longer in those with type 2 diabetes as the effect of weight loss may take longer to manifest in this patient group. It is important that physicians continue to monitor both body weight and the effects of treatment on risk factors during treatment with sibutramine and it should only be continued if demonstrable improvements in body weight and risk factors are observed. Because of concerns outlined above regarding the effects of sibutramine on blood pressure and heart rate it is important that this is monitored

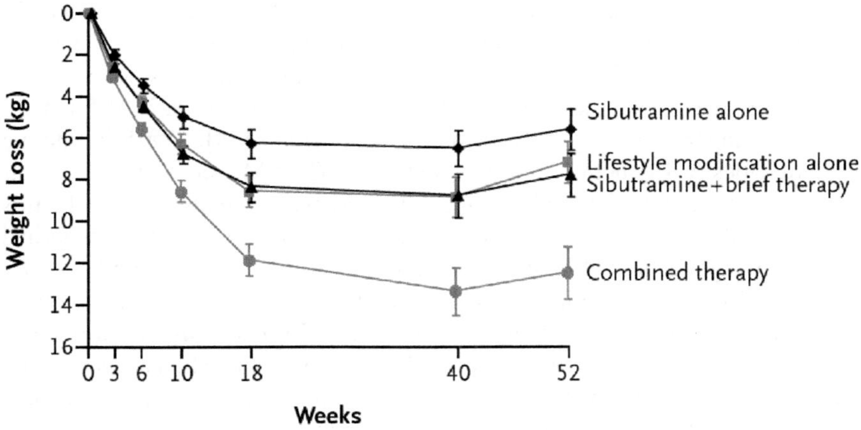

Figure 3. Effects of sibutramine with or without behaviour modification on body weight [22].

at least both before starting treatment and patients should be demonstrated to be normotensive (blood pressure less than 140/90) and not tachycardic (pulse less than 100 bpm). Pulse and blood pressure should be monitored every 2 weeks for the first 3 months of treatment, and periodically thereafter, sibutramine should be withdrawn if pulse or blood pressure rises by more than 10 mmHg or 10 bpm on two consecutive 2-weekly checks of pulse and blood pressure. The starting dose of sibutramine is 10 mg once daily. This can be increased to 15 mg daily if weight loss of 2 kg is not achieved after 1 month of treatment. Sibutramine is contraindicated in the presence of known organic causes of obesity, major eating disorders, major psychiatric illness, Tourette's syndrome, concomitant administration with monoabine oxidase inhibitors and other centrally acting drugs such as antidepressants and antipsychotic agents, the presence of severe cardiac disease and thyrotoxicosis and in severe hepatic and renal impairment (sibutramine summary of product characteristics, accessed online 17/07/07: http://emc.medicines.org.uk/emc/assets/c/html/displayDocPrinterFriendly.asp?documentid=14056). There is insufficient data to licence the use of sibutramine in children and young adults and in patients above 65 years of age, though emerging data may result in changes to these recommendations in the near future.

Efficacy in clinical trials

The effect of sibutramine on body weight is dose-related; in a 24 week dose ranging study, the mean weight loss from baseline was 1.2% with placebo, 2.7% with 1 mg of sibutramine, 3.9% with 5 mg, 6.1% with 10 mg, 7.4% with 15 mg, 8.8% with 20 mg and 9.4% with 30 mg sibutramine once daily [23].

Sibutramine has also been shown to increase weight loss and maintain weight loss, up to 1 year in patients who had previously lost weight using a very low calorie diet, with marked prevention of weight regain. In this study, 75% of subjects in the sibutramine group maintained a weight loss of at least 5% at 12 months, compared to only 42% in the placebo group [24]. The Sibutramine Trial of Obesity Reduction and Maintenance (STORM) study was a 2 year study of sibutramine for both weight loss and weight maintenance. In this study, all patients received active treatments for 6 months and those who had achieved weight loss of at least 5% of their body weight was assigned to either 10 mg of sibutramine or placebo. The dose of sibutramine could be increased up to 15 mg or 20 mg if weight regain occurred. In this study, those patients switched to placebo regained much of their lost weight, whereas those who continued sibutramine treatment were able to maintain weight loss with a weight regain of less than 2 kg over the 18 months of follow up. After 2 years treatment, 43% of patients had maintained 80% or more of their original weight loss, compared with only 16% of those treated with placebo (Fig. 4). This study also gave clear data regarding the effects of sibutramine on cardiovascular risk factors, with improvements on HDL cholesterol of 20.7% compared to 11.7% with placebo, and statistically significant falls in triglycerides, LDL cholesterol insulin, C-peptide and uric acid. Consistent with other studies, blood pressure rose in 3% of patients, with an average increase in systolic blood pressure of 0.1 mmol and diastolic blood pressure 2.3 mmHg, pulse by 4.1 bpm [25].

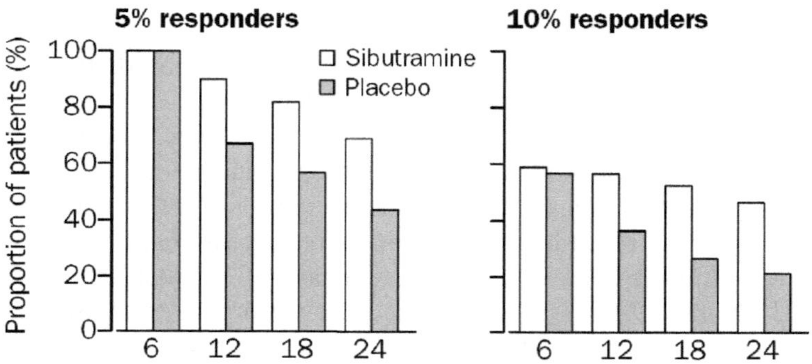

Figure 4. Proportion of patients reaching 5% and 10% weight loss in the STORM study [25].

Efficacy in obesity and co-morbidity

Type 2 diabetes

Sibutramine has been tested in a number of trials of patients with type 2 diabetes treated with a variety of agents, and in those who were treated with diet alone [26–29]. In these studies, the mean placebo subtracted weight loss was

4.5 kg and this tended to be associated with improvements in glycaemic control, reduction in fasting glucose of 0.4 mmol/l and mean reduction in HbA1c of about 0.7%; however the improvements in glycaemic control were greater in those who lost more weight, with as much as 1.1% fall in HbA1c in those patients who lost 10% of their body weight [30].

In sibutramine-treated patients with diabetes, improvements in HDL cholesterol and triglycerides were also seen compared to placebo, and the side-effect profile was similar as was seen in the non-diabetes studies.

Hypertension

Given the concerns about sibutramine and blood pressure, placebo control trials have been undertaken in patients with known hypertension, demonstrated that sibutramine was effective at producing weight loss in this group, and that blood pressure rises were modest, indeed those patients treated with placebo in these trials, and who were hypertensive were more likely to have a rise in blood pressure than those treated with sibutramine [31]. Effects on heart rate occurred independently of the rise in blood pressure.

Sibutramine in children and adolescents

Treatment of obesity in childhood remains a difficult area, and there is very limited information about what treatments are safe and clinically effective. It is also important to note that interpretation of weight loss and BMI change is more difficult to interpret in growing children than it is in adults. Three recent studies have investigated the use of sibutramine for up to 1 year in children and adolescents aged from 12–19 years [32–34]. These studies have in general shown that sibutramine is more effective than placebo in this age group, and that the effects are additive to behavioural strategies that form the basis of treatment. The largest study randomised 498 children aged 12–16 years to sibutramine 10 mg or placebo for 6 months, with the option of increasing the dose to 15 mg after 6 months if weight loss of 10% was not achieved. Overall weight loss was 8.4 kg greater in the sibutramine group compared to placebo, and the effects on cardiovascular risk factors such as HDL cholesterol, triglycerides and fasting insulin were similar to what has been seen in adults [33]. Nevertheless over one-third of patients did not complete the various studies, and in one of the smaller studies there were significant numbers of withdrawals (up to 10%) due to rises in pulse and blood pressure, although this was much lower in the other two published studies. Sibutramine may in the future have a role in the treatment of children and adolescents with obesity, but current data is limited in terms of numbers of patients studied, duration of follow up and long-term outcomes, so further studies in this area are needed before this can become a licensed indication.

Long-term outcomes of weight loss with sibutramine: The SCOUT study

As a result of concerns over cardiovascular effects of sibutramine, the European Medicines Evaluations Agency requested that the manufacturers conducted a large cardiovascular outcome study to evaluate the safety and efficacy of sibutramine in patients at high cardiovascular risk. The SCOUT trial is a randomised double blind comparison of sibutramine *versus* placebo, in addition to standard care for weight management, in overweight and obese subjects with an increased risk of cardiovascular disease. The study has recruited over 9,000 subjects who received treatment during a single blind lead in period with 10 mg of sibutramine, and who have been subsequently randomised to continue sibutramine, or switch to placebo treatment and who will be followed for at least 4 years.

The results of changes in body weight and blood pressure during this 6 week lead in period have recently been reported, along with the baseline characteristics of patients in the SCOUT study. Enrolled subjects came from one of three high risk cardiovascular categories; those with diabetes and at least one other risk factor; those with known cardiovascular disease, with at least one other risk factor but without type 2 diabetes; and those with both a previous history of cardiovascular disease, known type 2 diabetes and at least one other risk factor [19]. The latter group accounted for over two-thirds of patients in the study, with approximately 1,700 patients having cardiovascular disease only and 2,500 patients in the diabetes group. Mean weight loss during the 6 weeks run-in period was 2.2 kg and mean blood pressure actually fell by 3 mmHg diastolic and 1 mmHg diastolic with a 1.5 bpm rise in heart rate. Over 93% of enrolled patients completed this lead in period, and withdrawals due to adverse events amounted to 2.7% of withdrawals overall during the lead-in period. Relatively few patients were withdrawn due to sustained increases in blood pressure or heart rate and the overall death rate observed during the 6 week period was lower than expected from previous cardiovascular outcome studies, 1.2 per 100 years of exposure [35]. Clearly, it will be important and of considerable clinical interest to see the final results of the SCOUT study, which should report in 2009.

References

1 Buckett WR, Thomas PC, Luscombe GP (1988) The pharmacology of sibutramine hydrochloride (BTS 54 524), a new antidepressant which induces rapid noradrenergic down-regulation. *Prog Neuropsychopharmacol Biol Psychiatry* 12(5): 575–584
2 Weintraub M, Rubio A, Golik A, Byrne L, Scheinbaum ML (1991) Sibutramine in weight control: a dose-ranging, efficacy study. *Clin Pharmacol Ther* 50(3): 330–337
3 Luscombe GP, Slater NA, Lyons MB, Wynne RD, Scheinbaum ML, Buckett WR (1990) Effect on radiolabeled-monoamine uptake *in vitro* of plasma taken from healthy-volunteers administered the antidepressant sibutramine hcl. *Psychopharmacology* 100: 345–349
4 Luscombe GP, Hopcroft RH, Thomas PC, Buckett WR (1989) The contribution of metabolites to the rapid and potent down-regulation of rat cortical beta-adrenoceptors by the putative antide-

pressant sibutramine hydrochloride. *Neuropharmacology* 28(2): 129–134
5 Hind ID, Mangham JE, Ghani SP, Haddock RE, Garratt CJ, Jones RW (1999) Sibutramine pharmacokinetics in young and elderly healthy subjects. *Eur J Clin Pharmacol* 54: 847–849
6 Jackson HC, Needham AM, Hutchins LJ, Mazurkiewicz SE, Heal DJ (1997) Comparison of the effects of sibutramine and other monoamine reuptake inhibitors on food intake in the rat. *Brit J Pharmacol* 121: 1758–1762
7 Jackson HC, Hutchins LJ, Mazurkiewicz SE, Heal DJ, Buckett WR (1996) Comparison of the effects of sibutramine and other monoamine reuptake inhibitors on food intake in the rat. *Brit J Pharmacol* 117: 323
8 Halford JCG, Heal DJ, Blundell JE (1995) Effects in the rat of sibutramine on food-intake and the behavioral satiety sequence. *Brit J Pharmacol* 114: p387
9 Brown M, Bing C, King P, Pickavance L, Heal D, Wilding J (2001) Sibutramine reduces feeding, body fat and improves insulin resistance in dietary-obese male Wister rats independently of hypothalamic neuropeptide Y. *Brit J Pharmacol* 132(8): 1898–1904
10 Rolls BJ, Shide DJ, Thorwart ML, Ulbrecht JS (1998) Sibutramine reduces food intake in non-dieting women with obesity. *Obesity Res* 6: 1–11
11 Hansen DL, Toubro S, Stock MJ, Macdonald IA, Astrup A (1999) The effect of sibutramine on energy expenditure and appetite during chronic treatment without dietary restriction [Full text available]. *Int J Obesity* 23: 1016–1024
12 Barkeling B, Elfhag K, Rooth P, Rossner S (2003) Short-term effects of sibutramine (Reductil (TM)) on appetite and eating behaviour and the long-term therapeutic outcome. *Int J Obesity* 27(6): 693–700
13 Rowley HL, Butler S, Prow MR, Dykes SG, Aspley S, Kilpatrick IC, Heal DJ (2000) Comparison of the effects of sibutramine and other weight- modifying drugs on extracellular dopamine in the nucleus accumbens of freely moving rats. *Synapse* 38(2): 167–176
14 Cole JO, Levin A, Beake B, Kaiser PE, Scheinbaum ML (1998) Sibutramine: A new weight loss agent without evidence of the abuse potential associated with amphetamines. *J Clin Psychopharmacol* 18: 231–236
15 Connoley IP, Liu YL, Frost I, Reckless IP, Heal DJ, Stock MJ (1999) Thermogenic effects of sibutramine and its metabolites. *Brit J Pharmacol* 126: 1487–1495
16 Connoley IP, Frost I, Heal DJ, Stock MJ (1996) Role of beta-adrenoceptors in mediating the thermogenic effects of sibutramine. *Brit J Pharmacol* 117: p170
17 King DJ, Devaney N (1988) Clinical pharmacology of sibutramine hydrochloride (BTS 54524), a new antidepressant, in healthy volunteers. *Br J Clin Pharmacol* 26(5): 607–611
18 Birkenfeld AL, Schroeder C, Boschmann M, Tank J, Franke G, Luft FC, Biaggioni I, Sharma AM, Jordan J (2002) Paradoxical effect of sibutramine on autonomic cardiovascular regulation. *Circulation* 106(19): 2459–2465
19 James WPT (2005) The SCOUT study: risk-benefit profile of sibutramine in overweight high-risk cardiovascular patients. *Eur Heart J Supplements* 7(L): L44–L48
20 Finer N, Caterson I, Coutinho W, Van Gaal L, Maggioni A, Sharma A, Torp-Pedersen C, James WPT (2007) Clinically relevant weight loss achieved in high-risk patients during 6-week sibutramine treatment – an analysis from the sibutramine cardiovascular outcomes (SCOUT) trial. *Int J Obesity* 31: S29
21 Bray GA, Ryan DH, Gordon D, Heidingsfelder S, Cerise F, Wilson K (1996) Double-blind randomized placebo-controlled trial of sibutramine. *Obesity Res* 4: 263–270
22 Wadden TA, Berkowitz RI, Womble LG, Sarwer DB, Phelan S, Cato RK, Hesson LA, Osei SY, Kaplan R, Stunkard AJ (2005) Randomized trial of lifestyle modification and pharmacotherapy for obesity. *New Eng J Med* 353(20): 2111–2120
23 Bray GA, Ryan DH, Gordon D, Heidingsfelder S, Cerise F, Wilson K (1996) Double-blind randomized placebo-controlled trial of sibutramine. *Obesity Res* 4: 263–270
24 Apfelbaum M, Vague P, Ziegler O, Hanotin C, Thomas F, Leutenegger E (1999) Long-term maintenance of weight loss after a very-low-calorie diet: A randomized blinded trial of the efficacy and tolerability of sibutramine. *Am J Med* 106: 179–184
25 James WPT, Astrup A, Finer N, Hilsted J, Kopelman P, Rossner S, Saris WHM, Van Gaal LF (2000) Effect of sibutramine on weight maintenance after weight loss: a randomised trial. *Lancet* 356(9248): 2119–2125
26 Rissanen A, Taskinen MR (2000) Weight loss on sibutramine treatment for 12 months improves lipid profile in obese type 2 diabetic patients. *Diabetologia* 43: 657

27 McNulty SJ, Ur E, Williams G (2003) A randomized trial of sibutramine in the management of obese type 2 diabetic patients treated with mefformin. *Diabetes Care* 26(1): 125–131
28 Fujioka K, Seaton TB, Rowe E, Jelinek CA, Raskin P, Lebovitz HE, Weinstein SP (2000) Weight loss with sibutramine improves glycaemic control and other metabolic parameters in obese patients with type 2 diabetes mellitus. *Diabetes Obesity and Metabolism* 2(3): 175–187
29 Serrano-Rios M, Meichionda N, Moreno-Carretero E (2002) Role of sibutramine in the treatment of obese Type 2 diabetic patients receiving sulphonylurea therapy. *Diabetic Med* 19(2): 119–124
30 Norris SL, Zhang XP, Avenell A, Gregg E, Schmid CH, Kim C, Lau J (2004) Efficacy of pharmacotherapy for weight loss in adults with type 2 diabetes mellitus – A meta-analysis. *Arch Intern Med* 164(13): 1395–1404
31 Jordan J, Scholze J, Matiba B, Wirth A, Hauner H, Sharma AM (2005) Influence of Sibutramine on blood pressure: evidence from placebo-controlled trials. *Int J Obesity* 29(5): 509–516
32 Berkowitz RI, Wadden RA, Tershakovec AM, Cronquist JL (2003) Behavior therapy and sibutramine for the treatment of adolescent obesity – A randomized controlled trial. *JAMA* 289(14): 1805–1812
33 Berkowitz RI, Fujioka K, Daniels SR, Hoppin AG, Owen S, Perry AC, Sothern MS, Renz CL, Pirner MA, Walch JK et al. (2006) Effects of sibutramine treatment in obese adolescents – A randomized trial. *Ann Intern Med* 145(2): 81–90
34 Godoy-Matos A, Carraro L, Vieira A, Oliveira J, Guedes EP, Mattos L, Rangel C, Moreira RO, Coutinho W, Appolinario JC (2005) Treatment of obese adolescents with sibutramine: A randomized, double-blind, controlled study. *J Clin Endocrinol Metabolism* 90(3): 1460–1465
35 Torp-Pedersen C, Caterson I, Coutinho W, Finer N, Van Gaal L, Maggioni A, Sharma A, Brisco W, Deaton R, Shepherd G et al. (2007) Cardiovascular responses to weight management and sibutramine in high risk subjects: an analysis from the SCOUT trial. *Eur Heart J* doi:10.1093/eurheartj/ehm217

Pharmacotherapy of Obesity
Edited by J.P.H. Wilding

The endocannabinoid system as a target for obesity treatment

Muhammad Khan and John P.H. Wilding

Diabetes and Endocrinology Clinical Research Group, University Hospital Aintree, Clinical Sciences Centre, 3rd Floor, Lower Lane, Liverpool L9 7AL, UK

Introduction

During the past 30 years the prevalence of obesity has risen substantially in most developed countries. Across the world obesity is becoming one of the most preventable and modifiable metabolic disorders. There is evidence linking obesity to an increased risk of more than 30 medical conditions, raising an appropriate concern that this alarming trend will have major health consequences [1]. Serious conditions such as increased risk of type 2 diabetes mellitus (DM), coronary heart disease, hypertension, obstructive sleep apnoea and cancer, higher overall mortality rate and decreased lifespan have been associated with obesity [2, 3]. Morbid obesity can cause a decrease in life expectancy among young adults by as much as 5–20 years [4]. In addition to being a chief health concern, obesity is also becoming a major economic problem with significant consequences for health services worldwide. During the last 20 years beneficial trends have been evident in many cardiovascular disease risk factors including smoking, relative saturated fat intakes and cholesterol levels. Unfortunately, the parallel increase in adverse factors such as increased energy density of foods and reduced exercise, resulting in obesity, has counterbalanced or may have overwhelmed these benefits causing a growing prevalence of obesity-associated cardiometabolic disease [5]. The awareness of the health consequences of overweight and obesity, the benefits of modest weight loss and the frequent failure of lifestyle interventions for both weight loss and weight loss maintenance has led to the search for effective anti-obesity treatment. Evidence suggests that most obese individuals have an inappropriate control of their food intake rather than a metabolic defect in energy expenditure. This concept has turned attention toward drugs which reduce appetite or enhance satiety and so decrease energy intake as compared to thermoregulatory agents which increase energy expenditure.

The endocannabinoid system

Since the discovery of the first cannabinoid receptor together with its endogenous ligand in the early 1990s, the molecular basis for this novel neuromodulatory system has become better understood. The endocannabinoid system is now known to comprise a range of molecules, synthesised on demand from arachidonic acid precursors that regulate synaptic neurotransmission, together with their associated receptors [6]. This system acts as a neuromodulatory system affecting many physiological functions, not only in the central nervous system (CNS), but also in endocrine, reproductive, gastrointestinal, cardiovascular and immune systems. There are now known to be two types of endocannabinoid receptors, known as CB1 and CB2. Their natural ligands were identified as anandamide, monoacyl glycerol, 2-arachidonylglycerol and other fatty acid ethanolamides. The two most extensively studied endocannabinoids are anandamide and 2-arachidonylglycerol, both are synthesised from arachidonic acid and are lipid in nature. Interestingly, endocannabinoids are produced post-synaptically but act on pre-synaptic release of neurotransmitters, mainly causing inhibition of their release [6].

CB1 receptors are found in hypothalamus, amygdala, basal ganglia and cerebellum. In the central nervous system, by interacting with several of neuropeptides that modulate hunger and satiety signals the net result of activation of the endocannabinoid system is stimulation of appetite [8]. The evidence supporting this is described in more detail in Chapter 3 (Harrold and Wilding). Recent studies have shown that CB1 receptors are also located in peripheral tissues. Some CB1 receptors are found in gut and associated with its neural tissues, but others are located in adipose tissues, liver and muscle. CB1 receptors are also involved in the sterol regulatory element binding protein 1C (SREP-1C) pathway that is involved in regulator of hypothalamic – driven feeding behaviour through fatty acid synthetic FAS. In liver CB1 receptors tend to increase the production of VLDL via increase in acetyl CoA carboxylase 1 and fatty acid synthesis levels driven through SREBP-1C. This system also causes downregulations of LDL (apoE/B100) receptor in the liver [7].

Table 1. Overview of the endocannabinoid system

Site of action	Mechanism	Clinical implications
Adipose tissue	Decrease adiponectin Increase lipogenesis	Dyslipidaemia Insulin resistance
Liver	Increase fatty acid synthesis	Dyslipidaemia Insulin resistance Fatty liver
GI tract	Decrease satiety signals	Body weight
Muscle	Decrease insulin regulated glucose uptake	Insulin resistance
Hypothalamus Nucleus accumbens	Increase in food intake	Body weight

The CB2 system is mostly found in the immune system but is also expressed at low levels in the cerebral cortex, brainstem and in haemopoetic cells. The possibility exists that this system may also have a role in atherosclerosis as this process is closely related to immunity and inflammation.

The cumulative effect of blockage of central and peripherally located CB1 receptor may reduce motivation to eat palatable food (nucleus accumbens), anorexigenic effect (hypothalamus), stimulation of satiating signals engaging CB1 in several neural terminals (GI tract), increased adiponectin production, inhibition of lipogenesis (adipose tissue and liver) and increase glucose uptake (muscle) [8–10].

Rimonabant – a CB1 receptor antagonist

The understanding of endocannabinoid system has sparked the development of numerous CB1 receptor antagonists, of which rimonabant (SR 141716) is the first to complete clinical development and be available for use in patients.

Clinical pharmacology

Rimonabant (SR 141716) is a highly selective cannabinoid type CB1 receptor antagonist [11]. The chemical name is n-piperino-5-4 chlorophenyl)-1-(2,4 dichlorophenyl)-4-methylpyrazole-3-carboxamide (Fig. 1). Rimonabant specifically targets the endocannabinoid system where it shows high selectivity for the CB1 receptors while showing little affinity for CB2 receptors or other non-cannabinoid receptors. Rimonabant also binds to vanilloid receptor involved in some putative neuroprotective effects of this compound [12].

Rimonabant is a potent, selective and orally active cannabinoid antagonist with a long duration of action and shows high affinity for the predominantly

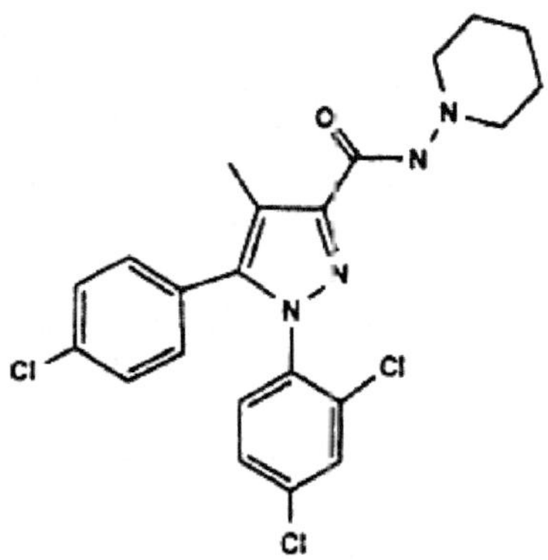

Figure 1. Structure of rimonabant.

centrally located CB1 cannabinoid receptor (Ki = 2 nm) and displays low affinity for the peripheral CB2 cannabinoid receptors (Ki > 1,000 nm) [13]. Maximum plasma concentrations of rimonabant are reached 1–3 h after its administration. Systemic exposure increases with increasing doses, although the increment is less than dose proportional. The plasma half life of rimonabant varies from 6–10 days in young adult subjects but tends to be longer in obese or elderly subjects, averaging up to 15 days in these individuals. Rimonabant is rapidly absorbed upon oral administration, with median Tmax (first time to reach maximum plasma concentration observed; max values in the range of 1–3.75 h that were generally independent of dose and dosing day. Exposure as assessed by Cmax, AVC 0–24 (area under plasma concentration curve from time zero to 24 h) and for AVC (area under plasma concentration curve extrapolated to infinity) increased with dose after single doses of 1 mg to 300 mg and repeated once daily doses of 3 mg to 60 mg in healthy patients. Rimonabant is well tolerated in up to 300 mg single doses and up to 60 mg daily given for 28 days. The primary pathway of metabolism is desterification. It is metabolised by cytochrome P450 3A4 (CYP3A4), therefore the CYP3A4 inhibitor (ketoconazole) causes a 104% (40–197%) rise in rimonabant concentrations. Secondary effects on rimonabant levels can also be seen with CYP2C8 inhibitors [7, 14].

Rimonabant is thought to regulate food intake via centrally mediated action [11, 15] but recent studies suggest it also has peripheral metabolic effects [16]. It has been suggested that cannabinoid receptors located on adipocytes might mediate some of these peripheral metabolic effects [17]. Rimonabant potentially blocks appetite for food and craving for drugs and has undergone clinical trials for obesity and smoking cessation. Essentially its target is to suppress signals in the brain, stomach and fatty tissues that promote smoking and eating.

Clinical trials in obesity

Rimonabant has recently undergone rigorous Phase III clinical trials to determine its potential in the management of obesity. These trials include RIO-North America, RIO-Europe, RIO-Diabetes and RIO-Lipids. The trials enrolled more than 6,600 patients worldwide and ranged from 1 to 2 years in length [18]. They were all prospective, randomised multicentred double-blind, placebo-controlled trials.

RIO-Europe and RIO-North America studied an obese or overweight population with or without co-morbidities, but excluding diabetes. RIO-Lipids studied an obese or overweight population with untreated dyslipidaemia while RIO-Diabetes examined an obese or overweight population with Type 2 diabetes. All four trials compared Rimonabant 5 and 20 mg/day with placebo. All subjects received diet and lifestyle therapy in addition to drug or placebo which is consistent with most pharmacological trials for weight reduction.

RIO-Lipids and RIO-Diabetes examined 1 year treatment, whereas RIO-Europe and RIO-North America were 2 year trials, the latter including an arm that was re-randomised to placebo or rimonabant after 12 months on active treatment. Each of the main studies is described in detail below, followed by a separate section on adverse events, as these were similar in all studies.

RIO-North America

Objectives and design

RIO-North America was a 2 year randomised double-blind, placebo-controlled trial. The main objective was to compare the efficacy and safety of rimonabant with placebo each in conjunction with diet and exercise for sustained changes in weight and cardio-metabolic risk factors over 2 years. A total of 3,045 patients were randomised in this study [19].

Principle findings

This study demonstrated that 20 mg/day of rimonabant in combination with standard dietary interventions produced greater weight loss in obese patients than placebo. The placebo-subtracted difference in body weight at one year was 4.7 kg using the last-observation carried forward analysis. There was some weight loss at the lower 5 mg dose, but this would not be considered adequate for registration of rimonabant for obesity management. Moreover the study showed beneficial changes in cardiovascular and metabolic risk factors over 1 year. Consistent with trials of other weight loss drugs, weight loss reached a plateau after about 9 months, but there was no weight regain in those patients treated for 2 years, whereas those reassigned to receive placebo in the second year regained most of the lost weight (Fig. 2). There were consistent differences from patients receiving placebo in the improvement of risk factors that are components of the metabolic syndrome, with favourable effects compared with placebo on fasting serum levels of triglycerides and HDL-cholesterol, blood pressure and waist circumference (Tab. 1). The prevalence of metabolic syndrome fell from 42% to 19.6%. One problem that has been highlighted in this trial and other trials of rimonabant is the high dropout rate (only 51% of patients completed 1 year of the study), nevertheless the effects on weight and metabolic parameters were robust to whichever statistical analysis was used to analyse the data, including the conservative baseline-imputed method.

RIO-Europe

The RIO-Europe trial was the second trial of rimonabant in uncomplicated obesity which assessed the efficacy and safety of rimonabant for weight reduction and its effects on metabolic risk factors [20].

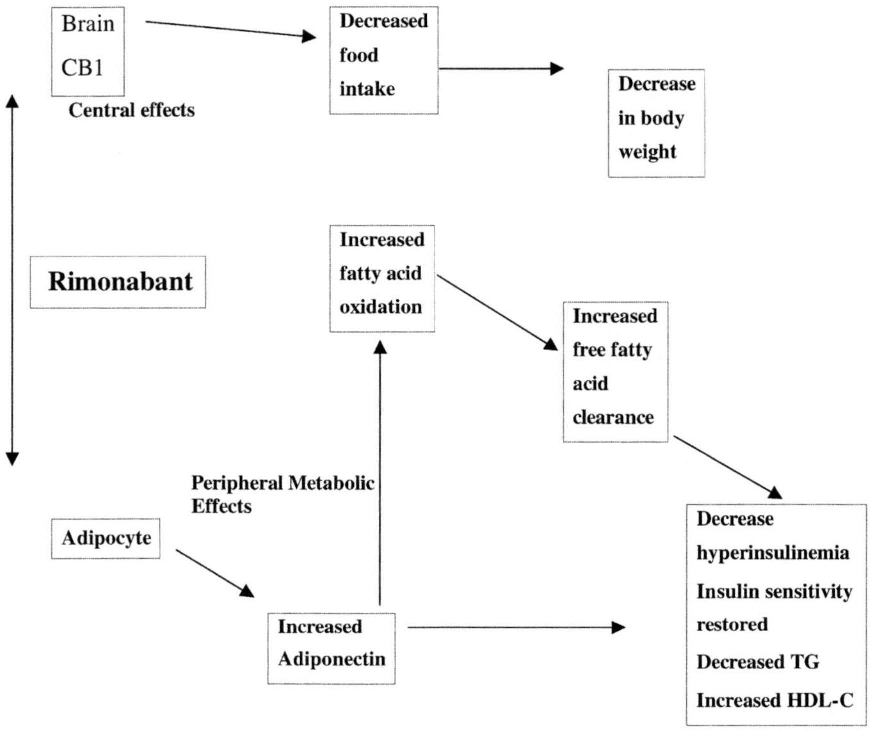

Figure 2. Rimonabant – effects on CNS and periphery in obesity.

RIO-Europe was also a Phase III, multinational multicentre, randomised, double-blind, and placebo-controlled trial comparing two fixed-dose regimens of rimonabant to placebo for a period of 2 years. The objectives of the trial were to assess the therapeutic effects of rimonabant on weight loss for a period of 1 year and to observe whether this drug potentially maintain weight loss during a second year of treatment or not. The study also evaluated the improvement in waist circumference and in metabolic risk factors such as glucose metabolism, dyslipidaemia and metabolic syndrome in conjunction with the safety and tolerability of the drug over the 2 year period.

RIO-Europe resulted in significant weight loss in subjects who were randomised to receive rimonabant treatment as compared to placebo. The results were very similar to those seen in RIO-North America. Although this trial was not specifically designed to assess people with diabetes, patients did show improvement in insulin sensitivity after rimonabant treatment. Finally, a key finding of this trial was that the prevalence of metabolic syndrome was reduced by more than half.

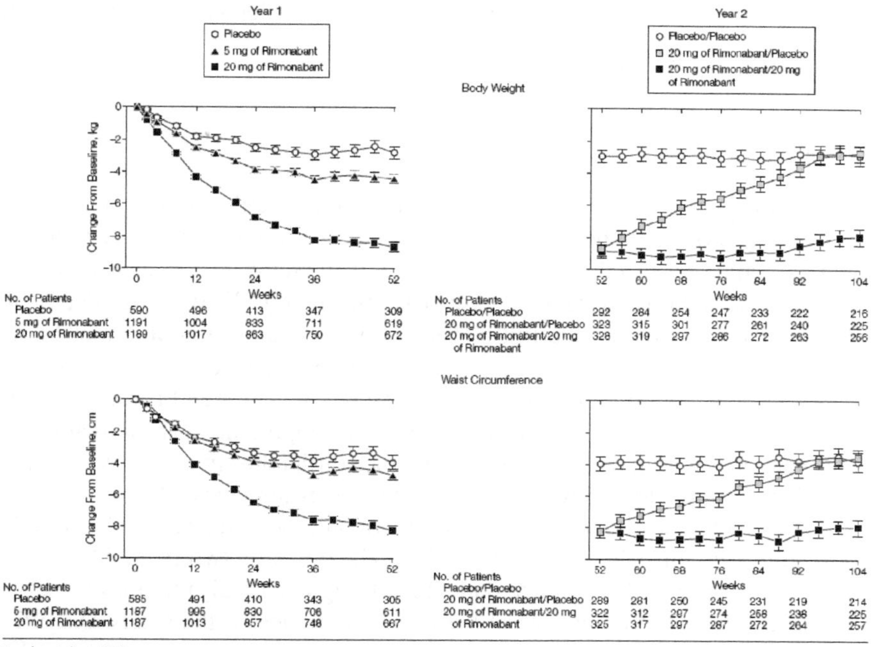

Error bars indicate SEM.

Figure 3. Weight and waist loss in the RIO North America study.

RIO-lipids

The Rimonabant In Obesity-lipid study evaluated the efficacy of rimonabant on metabolic risk factors, including adiponectin levels, in high risk patients who were obese and had dyslipidaemia. The primary objectives of the study was to assess the effect of 1 year of randomised double-blind treatment with rimonabant at a dose of 5 mg or 20 mg, as compared with placebo in addition to a hypocaloric diet (a deficit of 600 Kcal in relation to the calculated daily intake to maintain body weight), on the loss of body weight in patients who have untreated dyslipidaemia, are overweight or obese and do not have diabetes [21].

This study investigated the therapeutic use of rimonabant in high risk population of patients who are overweight or obese, with a focus on metabolic risk factors such as the size of LDL particles and levels of CRP and adiponectin. Treatment with rimonabant at a dose of 20 mg once daily as compared with placebo induced clinically significant weight loss and reduction in waist circumference suggesting marked mobilisation of abdominal fat, which by itself, would cause beneficial effects in cardiovascular risk profile.

Moreover the use of rimonabant also resulted in improvement in plasma triglycerides, total cholesterol, HDL-cholesterol level and HDL-cholesterol ratio, as well as changes in LDL particle size, glucose tolerance and adiponectin level.

RIO-Diabetes

The relation between Type 2 diabetes and most of the cardiovascular and metabolic risk factors including high blood pressure, high triglyceride concentration, low HDL-cholesterol and abdominal obesity is well established; diabetes is often considered as a cardiovascular disease risk equivalent [22]. It is regarded as a good clinical practice to give full consideration to treating multiple metabolic and cardiovascular risk factors in the management of Type 2 diabetes. Obese and overweight people are more prone to have Type 2 diabetes and cardiovascular disease, and weight loss is considered to be an important goal of treatment. However patients with diabetes often experience difficulties in losing weight and achieving desirable loss in weight in Type 2 diabetes remains a challenge.

RIO-Diabetes was a Phase III, multinational, randomised double-blind, multicentre, and placebo controlled trial which compared two fixed doses regimens of rimonabant (5 mg once daily and 20 mg once daily) to placebo for a period of 1 year. The objectives of the trial were to access the safety and efficacy of rimonabant in combination with exercise advice and a mild hypocaloric diet in overweight or obese patients with Type 2 diabetes who were already on monotherapy with sulfonylurea or metformin. The study was designed to investigate the effect of rimonabant on body weight, HbA_{1C} and other cardiometabolic risk factor in Type 2 diabetes as well as provide safety and tolerability of the drug in this population over 1 year [23].

The study reveals that patients treated with rimonabant 20 mg once daily benefited from greater weight loss than placebo (5.3 kg *versus* 1.4 kg). The overall weight loss was less than in the trials in uncomplicated obesity, but the placebo-subtracted weight loss was similar. In addition to weight loss, there was also a significantly greater fall in HbA_{1C}, reduced waist circumference, lower HDL-cholesterol, triglycerides and CRP concentration as well as systolic blood pressure in the rimonabant group. Interestingly, regression analysis suggested that the improvements in HbA_{1C} and HDL-cholesterol levels were twice that expected from the weight loss alone. These findings support the concept of peripheral metabolic effects of rimonabant [21].

Adverse effects of rimonabant

Given the widespread nature of the endocannabinoid system, it is perhaps surprising that adverse effects are largely limited to the gastrointestinal system and neuropsychiatric adverse effects. The most commonly reported adverse effects of rimonabant is nausea, an effect which tends to resolve over time, as witnessed by rates that are similar to placebo in the second year of the 2-year

Table 2. Summary of primary and secondary efficacy parameters, RIO studies

Efficacy Parameters	RIO-North-America	RIO-lipids	RIO-diabetes	RIO-Europe
Primary	Weight loss and weight maintenance over 1 year Prevention of weight regain during second year of treatment in obese patients with or or without co-morbidities	Weight loss and maintenance over 1 year	Weight loss and maintenance over 1 year	Weight loss and maintenance over 2 years (only 1 year data published)
Secondary	The effect over a period of 2 years on: – Dyslipidaemia – Hypertension – Quality of life – Tolerability and safety – Weight maintenance	The effect over a period of 1 year on: – Dyslipidaemia – Glucose tolerance status: rate of progression to the development of impaired glucose tolerance and type 2 diabetes, rate of improvement of glucose tolerance status – Hypertension – Fasting glycaemia and insulinaemia – Quality of life – Food intake and compliance to dietary prescription	The effect over a period of 1 year on: – Hypertension – Dyslipidaemia – Quality of life – Tolerability and safety – Glycaemic control	The effect over a period of 2 years on: – Dyslipidaemia – Glucose tolerance status: rate of deterioration of glucose tolerance status, rate of improvement of glucose tolerance status – Fasting blood glucose and insulin – Quality of life – Food intake and compliance to dietary prescription

studies, and relatively low withdrawal rates due to this side effect in the RIO trials.

Having the knowledge that cannabinoid CB1 receptors are located in certain areas of the brain which mediate stress responses (e.g., hypothalamus), and from preclinical data, it is expected that these receptors play a role in modulating emotional responses. Among the adverse effects that led to early withdrawal of patients, psychiatric events, mood changes in particular anxiety and depressed mood, accounted for about half of such drop outs. It is of interest that in those studies where hospital anxiety and depression scale was recorded, scores did not significantly change overall, perhaps suggesting there is a subgroup who are particularly susceptible to this adverse effect. However, as patients with current depression were excluded from the clinical studies, it will be important to monitor for this effect carefully as rimonabant is introduced into clinical practice.

Long-term outcome studies

A number of long-term outcome studies are either planned or underway with rimonabant; these include diabetes prevention studies (RAPSODI) and a cardiovascular outcome study (CRESCENDO).

Table 3. Rimonabant in four RCTs relative to placebo. Selected 1-year outcomes

Measure	RIO-lipids	RIO-diabetes	RIO-NA	RIO-Europe
Weight, kg	–5.4	–3.9	–4.7	–4.7
Systolic BP, mm Hg	–1.7	–2.3	–0.2	–1.2
Triglycerides % change	–12.4	–16.4	–13.2	–15.2
HDL-C, % change	8.1	8.4	7.2	8.9
Waist circumference, cm	–4.7	–3.3	3.6	–4.2

Table 4. Withdrawals due to psychiatric adverse events in RIO trials

Study	Placebo	Rimonabant 20 mgs
RIO-Europe	5.2%	7.0%
RIO-North America	2.3%	6.2%
RIO-lipids	2.3%	7.5%
RIO-diabetes	0.9%	3.6%

Table 5. Summary of studies in the obesity-related indications

Study	Treatment	% completed	Treatment duration	Diagnosis, inclusion criteria
RIO-Europe	Rimonabant 5 mg	47.76%	104 weeks	BMI > 30 kg/m or BMI > 27 with hypertension and/or dyslipidaemia.
	Rimonabant 20 mg	44.74%		Stable body weight.Type2 Diabetes excluded.
	Placebo	41.96%		
RIO-diabetes	Rimonabant 5 mg	64%	52 Weeks	BMI > 27 and <41.
	Rimonabant 20 mg	67.5%		Treated type2 diabetes.
	Placebo	66.3%		Stable body weight.
RIO-lipids	Rimonabant 5 mg		52 Weeks	BMI > 27 and <41.
	Rimonabant 20 mg			Type 2 diabetes excluded.
	Placebo			Untreated dyslipidaemia. Stable body weight
RIO-North-America Year 1	Rimonabant 5 mg	51.1%	52 Weeks	BMI > 30 kg/m or BMI > 27 with hypertension and/or dyslipidaemia. Stable body weight. Type 2 diabetes excluded
	Rimonabant 20 mg	55.2%	52 Weeks	patients from year 1 who achieved first year of treatment.
	Placebo	50.9%		
Year 2 (Re-randomisation)	Rimonabant 5 mg/placebo	70%		
	Rimonabant 5 mg/5 mg	71.6%		
	Rimonabant 20 mg/placebo	69.0%		
	Rimonabant 20 mg/20 mg	77.1%		
	Placebo/placebo	71.8%		

References

1 World Health Organisation (1998) *Obesity: preventing and managing the global epidemic*. Report of A WHO consultation on obesity, Geneva, 3–5 June 1997. World Health Organisation, Geneva
2 Allison DB, Fontaine KR, Manson JE, Stevens J, VanItallie TB (1999) Annual deaths attributable to obesity in the United States. *JAMA* 282(16): 1530–1538
3 Must A, Spadano J, Coakley EH, Field AE, Colditz G, Dietz WH (1999) The disease burden associated with overweight and obesity. *JAMA* 282(16): 1523–1529
4 Fontaine KR, Redden DT, Wang CX, Westfall AO, Allison DB (2003) Years of life lost due to obesity. *JAMA* 289(2): 187–193
5 Gregg EW, Cheng YJ, Cadwell BL, Imperatore G, Williams DE, Flegal KM, Narayan KM, Williamson DF (2005) Secular trends in cardiovascular disease risk factors according to body mass index in US adults. *JAMA* 293(15): 1868–1874
6 Matias I, Di Marzo V (2007) Endocannabinoids and the control of energy balance. *Trends Endocrinol Metabol* 18(1): 27–37
7 Wierzbicki AS (2006) Rimonabant: endocannabinoid inhibition for the metabolic syndrome. *Int J Clin Practice* 60(12): 1697–1706
8 Solinas M, Goldberg SR (2005) Motivational effects of cannabinoids and opioids on food reinforcement depend on simultaneous activation of cannabinoid and opioid systems. *Neuropsychopharmacology* 30(11): 2035–2045
9 Williams CM, Kirkham TC (1999) Anandamide induces overeating: mediation by central cannabinoid (CB1) receptors. *Psychopharmacology* (Berl) 143(3): 315–317
10 Engeli S, Bohnke J, Feldpausch M, Gorzelniak K, Janke J, Batkai S, Pacher P, Harvey-White J, Luft FC, Sharma AM et al. (2005) Activation of the peripheral endocannabinoid system in human obesity. *Diabetes* 54(10): 2838–2843
11 Colombo G, Agabio R, Diaz G, Lobina C, Reali R, Gessa GL (1998) Appetite suppression and weight loss after the cannabinoid antagonist SR 141716. *Life Sci* 63(8): L113–L117
12 Pertwee RG (2006) The pharmacology of cannabinoid receptors and their ligands: an overview. *Int J Obesity* 30: S13–S18
13 RinaldiCarmona M, Barth F, Heaulme M, Alonso R, Shire D, Congy C, Soubrie P, Breliere JC, LeFur G (1995) Biochemical and pharmacological characterization of Sr141716A, the first potent and selective brain cannabinoid receptor antagonist. *Life Sci* 56(23–24): 1941–1947
14 Henness S, Robinson DM, Lyseng-Williamson KA (2006) Rimonabant. *Drugs* 66(16): 2109–2119
15 Bellocchio L, Mancini G, Vicennati V, Pasquali R, Pagotto U (2006) Cannabinoid receptors as therapeutic targets for obesity and metabolic diseases. *Curr Op Pharmacol* 6(6): 586–591
16 Gary-Bobo M, Elachouri G, Scatton B, Le Fur G, Oury-Donat F, Bensaid M (2006) The cannabinoid CB1 receptor antagonist rimonabant (SR141716) inhibits cell proliferation and increases markers of adipocyte maturation in cultured mouse 3T3 F442A preadipocytes. *Mol Pharm* 69(2): 471–478
17 Van Gaal LF, Rissanen AM, Scheen AJ, Ziegler O, Rossner S (2005) Effects of the cannabinoid-1 receptor blocker rimonabant on weight reduction and cardiovascular risk factors in overweight patients: 1-year experience from the RIO-Europe study. *Lancet* 365(9468): 1389–1397
18 Pi-Sunyer X (2004) Clinical design of the studies of weight reduction and metabolic effects with rimonabant: The RIO (Rimonabant in Obesity) Program. *Obesity Res* 12: A27–A28
19 Pi-Sunyer F, Aronne LJ, Heshmati HM, Devin J, Rosenstock J (2006) Effect of rimonabant, a cannabinoid-1 receptor blocker, on weight and cardiometabolic risk factors in overweight or obese patients – RIO-North America: A randomized controlled trial. *JAMA* 295(7): 761–775
20 Van Gaal LF, Rissanen AM, Scheen AJ, Ziegler O, Rossner S (2005) Effects of the cannabinoid-1 receptor blocker rimonabant on weight reduction and cardiovascular risk in overweight patients: 1-year experience from the RIO-Europe study (vol 365, pg 1389, 2005) *Lancet* 366(9483): 370
21 Despres JP, Golay A, Sjostrom L (2005) Effects of rimonabant on metabolic risk factors in overweight patients with dyslipidemia. *N Eng J Med* 353(20): 2121–2134
22 Haffner SM, Lehto S, Ronnemaa T, Pyorala K, Laakso M (1998) Mortality from coronary heart disease in subjects with type 2 diabetes and in nondiabetic subjects with and without prior myocardial infarction. *N Engl J Med* 339(4): 229–234
23 Scheen AJ, Finer N, Hollander P, Jensen MD, Van Gaal LF (2006) Efficacy and tolerability of rimonabant in overweight or obese patients with type 2 diabetes: a randomised controlled study (vol 368, pg 1660, 2006) *Lancet* 368 (9548): 1650

Pharmacotherapy of Obesity
Edited by J.P.H. Wilding

Using the body's natural signals – gut hormones

Owais B. Chaudhri, Kirsty L. Smith and Stephen R. Bloom

Department of Metabolic Medicine, Imperial College London, Hammersmith Hospital, London W12 0NN, UK

Introduction

The balance between energy (food) intake and energy expenditure is closely regulated to keep body weight stable over time. Increasingly, however, the homeostatic mechanisms responsible are failing to keep pace with societal changes in eating behaviour and activity levels. The prevalence of obesity in many parts of the world has now reached epidemic proportions. It is estimated that it causes 30,000 deaths per year in the UK, and in the USA it is set to overtake smoking as the leading cause of preventable illness and premature death [1, 2]. Current strategies for the non-surgical treatment of the morbidly obese have met with limited success [3–5] and without the development of more effective treatments, the socioeconomic and public health implications of an unchecked rise in obesity are grave.

Gut hormones are important physiological mediators of appetite regulation [6, 7] and therefore these hormones, their receptors and the pathways through which they act offer novel targets for the development of effective therapies for obesity. Here we review the principles upon which a gut hormone-based treatment may be based, as well as consider possible limitations associated with the use of particular peptides.

General principles

The gastrointestinal tract is the largest endocrine organ in the human body. The array of known gut hormones is extensive and their primary physiological role is to optimise the function of the gut as the organ of digestion and nutrient absorption [8]. In this regard, in the century since the description of the first gut hormones, investigators initially focused on the actions of gut hormones on aspects of gastrointestinal function. Postprandial gut hormone secretion into the bloodstream leads to changes in gastrointestinal motility and exocrine secretion [8–11]. Thus, by regulating the delivery of nutrients to specific compartments of the alimentary canal, gut hormones contribute significantly to efficient digestion.

Control of appetite arguably constitutes another point at which nutrient delivery to the gut may be influenced. In recent decades, it has come to be realised that gut hormones also influence food intake and they now occupy a central place in the complex neuroendocrine interactions that underlie appetite control (Fig. 1) [6, 7]. The majority of gut hormones that have been found to alter food intake promote satiety. The exception to this is the hormone ghrelin, which increases hunger and has been implicated in meal initiation. The development of therapies based on gut hormones for the treatment of obesity carries with it the advantage that in manipulating the pathways by which appetite is regulated under normal physiological conditions, the likelihood and severity of side effects is lessened. This contrasts, for example, with intervention at the level of central nervous system (CNS) appetite circuits. Many of the neurotransmitters involved in regulation of energy balance, such as neuropeptide Y (NPY) and γ-amino butyric acid (GABA), are ubiquitous and perform numerous other functions within the CNS. Direct manipulation of neuronal circuits therefore risks disruption of unrelated physiological functions.

Possible strategies for the development of gut hormone-based therapies include the use of receptor agonists, long-acting gut hormone analogues, agents that extend endogenous gut hormone half-life by inhibiting breakdown, and the enhancement of endogenous release. Tantalisingly, gut hormone levels are altered following gastrointestinal surgery for obesity, and this may be one mechanism by which weight loss seen following these procedures occurs [12–14]. Thus far, however, the non-surgical enhancement of endoge-

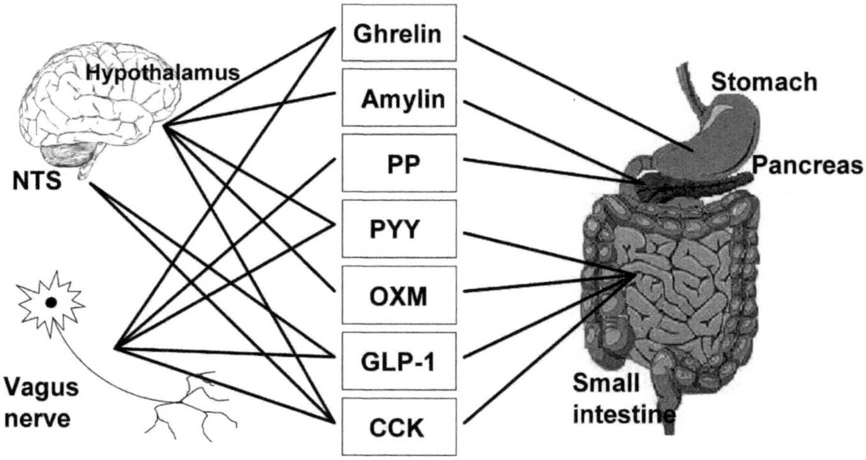

Figure 1. The gut hormones affecting food intake. The gut hormones are secreted from different organs within the alimentary canal including the stomach, the pancreas and the small intestine. The gut hormones' signals are conveyed to the centres for the control of appetite either directly via the hypothalamus or indirectly via the brainstem or the vagus nerve. CCK – cholesystokinin; GLP-1 – glucagon like peptide 1; NTS – nucleus of the solitary tract; OXM – oxyntomodulin; PP – pancreatic polypeptide; PYY – peptide YY.

nous gut hormone release is an area largely unexplored and the major focus of current research is the development of gut hormone analogues and receptor agonists.

Central circuits of appetite control

A comprehensive review of the neuronal centres of appetite regulation is beyond the scope of this article, and the interested reader is directed elsewhere [15, 16]. Briefly, however, the hypothalamus and dorsal vagal complex are major centres of the control of energy balance. The arcuate nucleus (ARC) of the hypothalamus acts as a conduit and co-ordinates the integration of the neuronal and humoral aspects of control of food intake. Many gut hormones influence neuronal activity within the ARC indirectly, via actions on the vagus nerve and dorsal vagal complex, which communicates directly with the ARC [6, 7]. Neurones within the ARC also express receptors for gut hormones and some hormones may act there directly, although this remains controversial. The ARC contains two important populations of neurones involved in the regulation of food intake. One population, which co-expresses the neuropeptides NPY and agouti-related protein (AgRP), causes a stimulation in food intake while the second co-expressing pro-opiomelanocortin (POMC) and cocaine- and amphetamine-regulated transcript (CART) is anorexigenic in activity [15, 16].

Gut hormones that signal satiety

Cholecystokinin

The peptide cholecystokinin (CCK) is both a gut hormone and a neurotransmitter. Two G-protein-coupled receptors for CCK are known, designated CCK1R and CCK2R (formerly CCK-A and CCK-B receptors, respectively) [10]. Released postprandially from I-cells of the small intestine, CCK has well-documented effects on gut motility and secretion, but also inhibits food intake in both rodents [17] and humans [18]. This effect is thought to be mediated via the CCK1R. Chronic administration of CCK1R antagonists or anti-CCK antibodies results in weight-gain [19, 20] and lack of the CCK1R, as in the OLETF rat, is associated with hyperphagia and obesity [21].

This perturbation of normal physiology in the absence of CCK signalling argues for a critical role for CCK in the control of food intake. The potential for capitalising on this in developing an effective therapy for obesity is tempered, however, by the observation that repeated administration of CCK does not result in loss of weight. Although meal size is reduced, there is a compensatory increase in meal frequency [22, 23]. Moreover, continuous administration of CCK results in loss of its anorectic effects after 24 h [24]. GlaxoSmithkline

(Brentford, Middlesex, UK) recently halted trials of its CCK1R agonist 181771 after results made it commercially non-viable.

PP-fold proteins

The PP-fold family of proteins consists of NPY, peptide YY (PYY) and pancreatic polypeptide (PP). They are all 36 amino acids in length and share a common U-shaped structural motif known as the PP-fold [25]. All three influence appetite, NPY as a neurotransmitter acting within the CNS, and both PYY and PP as hormones released by the gastro-entero-pancreatic system in response to nutrient ingestion. All three also share a receptor family. Five receptors have been cloned in mammals thus far, Y_1, Y_2, Y_4, Y_5 and y_6, although the last of these is truncated and non-functional in humans [26]. The receptors are classified according to their affinity for different ligands, and it is this diversity, as well as differences in receptor distribution and coupling to downstream signalling pathways, that allows the PP-fold family of proteins to mediate a wide range of functions throughout the body (Tab. 1).

In particular, Y_1 and Y_5 receptors are distributed throughout important appetite-regulatory regions of the brain, including the hypothalamic paraven-

Table 1. The Y receptor family

Receptor	Ligand affinity	Distribution
Y_1	NPY = PYY > PP	Dorsal root ganglia Amygdala Hypothalamus Blood vessels Cortex
Y_2	$NPY = PYY = PYY_{3-36} > PP$	Hypothalamus Dorsal root ganglia Hippocampus Intestine
Y_4	$PP > PYY \geq NPY > PYY_{3-36}$	Hypothalamus Amygdala Thalamus Intestine Pancreas Heart Muscle
Y_5	NPY = PYY > PP	Hypothalamus Thalamus Nucleus of the solitary tract

The distribution of the Y receptors and the affinities for the ligands which bind them. To date five receptors have been cloned in mammals although the fifth (y_6) is truncated and non-functional in humans and therefore is not described here.

tricular nucleus (PVN), where they mediate the orexigenic actions of NPY. Y_2 receptors, however, for which the cleavage product PYY_{3-36} demonstrates relative specificity, are found in high concentration in the ARC. Here they are thought to act as presynaptic inhibitory autoreceptors, inhibiting the release of NPY, and thus reducing food intake.

Pancreatic polypeptide

PP is principally released by cells located in the periphery of the pancreatic islets, although a small amount is produced by the exocrine pancreas and cells of the intestinal mucosa. It is released into the circulation postprandially in proportion to the calories ingested [27].

The place of PP in gastrointestinal physiology is a matter of ongoing debate. This extends to its effects on food intake. A number of investigators have shown an inhibitory effect of PP on food intake in rodents and man [28–30]. Further, circumstantial, evidence for a role in appetite regulation takes the form of a blunted PP response in hyperphagic, obese children with Prader-Willi syndrome (PWS) [31] and an exaggerated response in anorexic individuals [32]. Genetically obese *ob/ob* mice lack PP cells in the pancreas and replacement with twice-daily intraperitoneal injections of bovine PP results in a reduction in body weight gain [33]. This, coupled with the observation that chronic over-expression of PP in rodents promotes weight loss [34], makes a therapy based on PP a promising prospect.

PP shows relative specificity for the Y_4 receptor subtype [26], which is known to be distributed within the hypothalamus, but also within the dorsal vagal complex in the brainstem. Injection of PP directly into the ARC actually promotes feeding [35]. This raises the possibility that PP exerts its anorexigenic effect by acting at peripheral sites, such as the vagus, rather than by direct interaction with CNS centres of appetite control. Thus, a treatment based on PP might be expected to have little CNS involvement, which again increases its potential utility.

There are a number of difficulties with this approach, however. PWS does not constitute a good model for the majority of obesity, which is non-syndromic in nature and multifactorial in aetiology. When PP responses are examined in subjects with simple obesity, no significant blunting is observed [31, 36, 37] and other investigators have failed to reproduce the appetite-inhibitory effects of exogenous PP in lean subjects [38]. Even in subjects with PWS, the effects of infused PP are equivocal [39]. While a treatment for obesity based on PP is not an impossibility, there are clearly hurdles to be overcome.

Peptide YY

PYY is synthesised in the L-cells of the gastrointestinal tract. Levels of PYY in the gut are lowest in the proximal small intestine, and progressively rise towards the distal gastrointestinal tract to reach a peak in the rectum [40]. In common with other appetite-inhibiting gut hormones, PYY is released into the circulation in response to a meal in proportion to the calories ingested [40].

It has been known for some time that PYY modifies food intake, although the initial studies involving CNS administration indicated the peptide to be a mediator of orexigenic behaviour [41]. There are two circulating forms of PYY, the full-length peptide and an N-terminal truncated form, PYY_{3-36}, which is the major circulating form. This species of peptide, injected directly into the cerebroventricular system, also promotes feeding [42].

Peripheral administration: Batterham and co-workers, however, investigated the effects of PYY_{3-36} injected peripherally into rodents, arguably a better model for the physiological release of PYY_{3-36} after a meal, and found it to reduce food intake. This inhibition of feeding was evident at a dose which gave plasma levels concordant with those seen in the normal postprandial state [43]. Following infusion to lean human subjects, PYY_{3-36} also reduced energy intake and appetite scores, again at plasma levels similar to those seen physiologically [43]

The apparent dichotomy in the actions of peripherally-administered *versus* CNS-injected PYY_{3-36} is possibly due to differences in Y receptor subtype distribution, as discussed above. The orexigenic effects of intracerebroventricular (ICV) PYY_{3-36} are attenuated in Y_1 and Y_5 receptor-knockout mice, suggesting that ICV PYY_{3-36} is acting via these receptors [42]. The inhibitory effects of PYY_{3-36} on appetite, however, are mediated via Y_2 receptors [43, 44], which are localised particularly on NPY neurones in the ARC. Inhibition of NPY release by Y_2 receptor activation might therefore reduce food intake.

Thus, it is argued, ICV PYY_{3-36} has access to receptors in deep hypothalamic nuclei such as the PVN, and causes an increase in food intake. The ARC, on the other hand, is anatomically close to the median eminence, which lacks a complete blood brain barrier. Batterham et al. proposed that peripherally-injected PYY_{3-36} preferentially accesses Y_2 receptors in the ARC to decrease food intake [43]. The corollary of this is that injection of PYY_{3-36} directly into the ARC, contrasting with less-targeted ICV administration, results in an inhibition of feeding, and the actions of peripheral PYY_{3-36} on feeding are lost in Y_2-null mice [43]. Similarly, specific antagonism of the Y_2 receptor inhibits PYY_{3-36}-induced hypophagia [44, 45].

The 'leaky ARC' model of the actions of circulating PYY_{3-36} remains controversial, however. PYY_{3-36} is able to cross the blood-brain barrier freely by non-saturable means [46], and this casts doubt on the significance of the anatomical proximity of the ARC and median eminence. Additionally, there is a growing body of evidence that the effects of PYY_{3-36} on appetite, and indeed its induction of c-fos within the ARC, are mediated via the vagus nerve [47, 48]. This latter arrangement echoes the mechanism of action of a number of other gut hormones (such as CCK and glucagon-like peptide 1 (GLP-1)) and from a therapeutic perspective, offers an advantage in the form of the ability to manipulate the effects of PYY_{3-36} on food intake at a peripheral site, away from the CNS. Recent evidence that vagotomy or area postrema ablation actually potentiates the anorexigenic actions of peripheral PYY_{3-36} has added a further layer of complexity to the work of characterising the pathways through which PYY_{3-36} acts [49, 50].

Batterham et al. also demonstrated another characteristic of PYY_{3-36} that casts a favourable light on its potential for therapeutic application. Despite lower basal levels of PYY_{3-36} and a blunted postprandial response in obese humans, obesity does not appear to be associated with resistance to the effects of PYY_{3-36} [51, 52]. This contrasts with other peripheral signals of energy balance, such as leptin.

Variability of effect in rodent models: There are, however, substantial obstacles to the development of an effective PYY_{3-36}-based therapy for obesity. Not least of these is a *prima facie* capriciousness in the ability of peripherally-administered PYY_{3-36} to suppress food intake. The difficulties that some investigators have experienced in reproducing this effect extend across a variety of rodent strains and a number of experimental paradigms (reviewed in [53])

Studies supportive of the original work by Batterham et al. have also been published, however. These too extend across different rodent species and strains, although there is some variability between studies in the duration of the hypophagia [43, 46, 50, 54–65]. The reasons for the apparent variability in the capacity of PYY_{3-36} to inhibit feeding remain obscure, although a number of possibilities have been posited. One explanation centres on the observation that stress potently inhibits feeding [59, 66, 67]. It has been argued that animals that have not been habituated to the experimental procedure and are therefore stressed during the experiment itself will naturally eat less than well-acclimatised animals. This may mask the anorectic effect of the peptide being studied. Despite increased awareness about the importance of animal acclimatisation, some laboratories remain unable to demonstrate an inhibitory effect of PYY_{3-36} of food intake [53], suggesting that other, as yet unidentified factors are also operating.

Recent data have indicated a possible mechanism that might account for some of this variability. PYY_{3-36} modulates not only NPY release in the hypothalamus, but also acts on the anorexigenic POMC system. Differences in experimental procedure and habituation technique might make animals more or less sensitive to the effects of stressors on food intake [68, 69]. Differences in infusion protocol, and the length of time for which plasma levels of PYY_{3-36} are elevated has also been invoked as an explanation [49].

Arguments that an effect of stress on the effectiveness of PYY_{3-36} diminishes its usefulness as a basis for therapy are not necessarily applicable to humans, in whom higher neurological centres also have a significant influence on feeding behaviour. In a move away from problematic rodent models, some authors have elected to investigate the acute effects of PYY_{3-36} on food intake in non-human primates. Arguably, this constitutes a better model for human physiology. Moran and colleagues demonstrated a reduction in food intake following PYY_{3-36} administration to rhesus monkeys [70], mirroring the effect of acute infusion of the peptide into humans [43, 51].

Chronic administration: Two, interconnected, requirements of any therapy for obesity are that the treatment must bring about a reduction in body weight, and that repeated application of the treatment should not result in a significant loss of this effect over time. Here, too, there is some controversy in the literature regarding the viability of PYY_{3-36} as a potential therapy. While Batterham et al. showed that twice-daily administration of PYY_{3-36} to rats for 7 days resulted in a reduction in body weight gain [43], others have again found this a difficult finding to replicate [53]

Pittner et al., however, did report a decrease in body weight in a number of rodent models of obesity following chronic, continuous administration of the peptide via a pump [64]. Similarly, recent data demonstrate a significant reduction in weight and a trend towards lower food intake in New Zealand white rabbits following once-daily injection of high dose PYY_{3-36} [71]. A study conducted in rhesus monkeys also confirmed a reduction in weight following administration of PYY_{3-36} twice-daily for 2 weeks [72], although this result is tempered by a number of caveats of which the most significant is that supraphysiological plasma levels of PYY_{3-36} were required. No behavioural patterns suggestive of excess nausea were observed, however.

Therapeutic implications: The difficulties associated with the use of PYY_{3-36} in rodent models introduce a level of complexity to the process of translating the original findings of Batterham et al. into a therapy for obesity. However, whether such a treatment will be useful in a clinical context is dependent ultimately on the effect of the peptide in humans. The findings that PYY_{3-36} reduces food intake in lean humans, and that it is just as efficacious acutely in obese volunteers [43, 51], count in its favour. Two pharmaceutical companies are currently developing PYY_{3-36}-based treatments for obesity

Nastech Pharmaceutical Company Inc (Bothell, WA), in collaboration with Merck & Company Inc (Whitehouse Station, NJ), has recently completed Phase Ic trials of PYY_{3-36} delivered via the intranasal route. Acutely, PYY_{3-36} caused significant reductions in visual analogue appetite scores, and there was a trend towards a reduction in food intake at a test meal [73]. The most significant adverse effect noted was nausea, but this was seen in those subjects with the highest plasma levels of PYY_{3-36} following administration. Furthermore, preprandial use of the nasal spray for 6 days in 37 obese volunteers was associated with significant reductions in daily caloric intake that were sustained over the study period. Thrice-daily administration of the peptide yielded a reduction in caloric intake of 648 calories (2713 kJ) and a reduction in body weight of 1.3 pounds (0.6 kg) after 6 days of treatment.

Amylin Pharmaceuticals Inc (San Diego, CA) is also engaged in the development of PYY_{3-36} for use as a treatment for obesity. Their investigational compound AC162352 has completed Phase I clinical trials, although no data are currently in the public domain regarding its efficacy. Thus, while undeniably challenging, the translation of the observations of Batterham et al. and others into a useful clinical therapy is not unfeasible. Further, larger studies in which the peptide is administered to volunteers for longer periods are required and the results of these are awaited with interest.

Products of proglucagon cleavage

Proglucagon is a 160-amino acid prohormone that is synthesised in the α-cells of the pancreatic islets, the L-cells of the gastrointestinal tract and within the CNS [74]. Proglucagon undergoes tissue-specific cleavage by prohormone convertases 1 and 2 to result in a number of biologically-active fragments (Fig. 2). Within the CNS and the L-cells of the gut, the products glucagon-like peptide 1 (GLP-1) and oxyntomodulin (OXM) have been implicated in the regulation of appetite and energy intake.

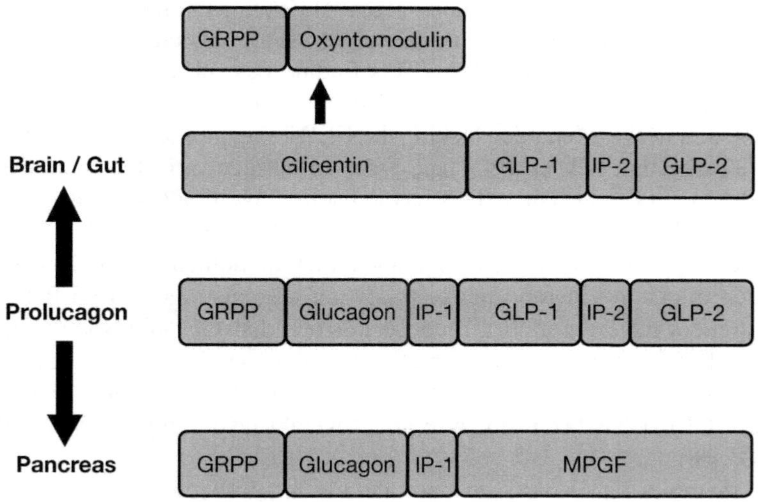

Figure 2. Proglucagon processing. The structure of mammalian proglucagon. The products of proglucagon are subject to tissue specific processing by prohormone convertase enzymes. GLP – glucagon like peptide; GRPP – glicentin related pancreatic polypeptide; IP – intervening peptide; MPGF – major proglucagon fragment.

Glucagon-like peptide 1

GLP-1 co-localises with OXM and PYY in the L-cells of the gastrointestinal tract [75]. Like PYY, GLP-1 exists in the circulation in a number of forms. GLP-1 may be cleaved from its prohormone as either a 36- or 37-amino acid molecule, depending on whether the C-terminal glycine is present. Further truncation by removal of the six N-terminal amino acids is required for biological activity. Both peptide species (GLP-1_{7-37} and GLP-$1_{7-36amide}$) are equipotent at eliciting all known actions of GLP-1, although GLP-$1_{7-36amide}$ is the predominant circulating form [76].

The effect of GLP-1 that has attracted most interest from a therapeutic viewpoint is its potent action as an incretin and in keeping with this, GLP-1 is released into the circulation after a meal in proportion to the calories ingested [77]. Lately, however, increasing research has been directed at determining the role of GLP-1 in appetite-regulation.

The GLP-1 receptor is distributed widely, and this reflects the varied actions of GLP-1, both as a hormone, and within the CNS as a neurotransmitter. Binding studies and RT-PCR data have localised the receptor to a number of areas of the CNS key to the regulation of energy balance, including the ARC and the PVN in the hypothalamus, and the area postrema of the brainstem [78, 79]. Turton et al. injected GLP-$1_{7-36amide}$ ICV into rats and found that it significantly inhibited food intake. Importantly, the GLP-1 receptor antagonist exendin$_{9-39}$ was found to have no effect on fasted rats, but powerfully increased feeding in satiated rats, implying a role for GLP-1 in the physiological regulation of food intake [80]. Furthermore, repeated ICV injection of GLP-$1_{7-36amide}$ over 6 days resulted in a loss of body weight. Again, injection of exendin$_{9-39}$ resulted in the opposite effect [81]. This finding has subsequently been confirmed and extended to peripheral administration of the peptide [77]. It has been noted, however, that mice lacking the GLP-1 receptor are diabetic, but not obese [82] and that ICV GLP-$1_{7-36amide}$ is a strong inducer of the phenomenon of conditioned taste aversion. This has led some investigators to question the role of GLP-1 in the normal physiology of appetite regulation [83].

A recent meta-analysis has concluded that in humans there is a dose-dependent reduction of food intake following administration of GLP-1 [84]. Some authors have reported a reduced postprandial GLP-1 response in the obese [85] which would make it an attractive therapeutic candidate, although other investigators have disputed these findings [86]. Nevertheless, administration of GLP-1 has been shown to affect weight favourably in obese volunteers [87, 88].

Two therapies based on GLP-1 are already in advanced stages of development, ostensibly as adjunctive therapies for the treatment of type 2 diabetic patients. Each adopts a slightly different strategy in order to overcome the greatest difficulty in the use of native GLP-1, namely that the peptide undergoes rapid degradation in the circulation by the enzyme dipeptidyl peptidase IV (DPP-IV) [89].

Amylin Pharmaceutical Company Inc has recently been granted FDA and European licensing approval for the use of its drug exenatide in poorly controlled diabetes. A synthetic form of exendin-4, a naturally-occurring peptide found in the venom of the gila monster, *Heloderma suspectum*, exendin-4 is a DPP-IV-resistant agonist at the GLP-1 receptor. The benefits of a therapy for type 2 diabetic patients that improves glycaemic control while also promoting weight loss are obvious [90].

The GLP-1 analogue Liraglutide, under development by Novo Nordisk (Copenhagen, Denmark), avoids degradation by DPP-IV by a single amino acid substitution and the attachment of an acyl side chain that allows for noncovalent binding to albumin [89]. It is currently in Phase II trials, although specifically with respect to its use in obesity, twice-daily injection for 10 days has been shown to induce weight loss in both normal rats and in rats made obese by neonatal exposure to monosodium glutamate [91]. Human data demonstrating a reduction in weight after 5 weeks' administration have also been published [89].

When given to non-diabetic subjects neither peptide appears to cause significant hypoglycaemia, and so their utility may extend beyond patients with diabetes. Both are, however, associated with a significant degree of nausea, which is dose-dependent [89], although this side effect may be ameliorated by gradual dose-escalation on institution of therapy [89, 92]. Another concern relating specifically to exenatide is the potential for patients to develop antibodies against the foreign peptide. Reportedly, up to 30% of patients have antibodies detectable after prolonged administration, although thus far, this has not progressed to significant clinical effects [89]. GLP-1 receptor stimulation is associated with an increase in blood pressure and heart rate [93, 94]. The effect of long-term administration of GLP-1 receptor agonists on cardiovascular parameters, particularly in a patient population that has a significant baseline risk of cardiovascular disease, remains to be evaluated.

Oxyntomodulin

Co-secreted from gastrointestinal L-cells with PYY and GLP-1, OXM was named for its ability to reduce gastric acid production. In both rodents and humans, OXM reduces energy intake [95–99]. The inhibition of this effect by the GLP-1 receptor antagonist exendin$_{9-39}$ and the loss of effect in GLP-1 receptor-deficient mice suggest that OXM is acting via the GLP-1 receptor [96, 97]. The affinity of OXM for the GLP-1 receptor, however, is two orders of magnitude lower than that of GLP-1, and this has led to the suggestion that OXM may be acting via a different receptor system [96].

Chronic administration of the peptide has also been shown to result in weight loss. In a recent study, OXM was administered by subcutaneous injection preprandially into obese human subjects over a period of 4 weeks. This resulted in a significant reduction of body weight of 2.3 kg in the treatment group, compared with 0.5 kg in the control [99]. Although direct comparisons have not been made, OXM appears to cause fewer gastrointestinal side effects than GLP-1-based treatments and this makes it a potentially propitious contender as the basis for an effective anti-obesity therapy. Part of the explanation for this may lie in the observation that as well as reducing food intake, OXM may also affect body weight through an increase in energy expenditure. Supporting this model, is the observation that rats repeatedly injected with OXM ICV for 7 days gained less weight than pair-fed controls [100]. Core temperature was also higher and brown and white adipose tissue mass were reduced in OXM-injected rats, compared with controls.

The translation of these encouraging early results into a useful therapy, however, awaits the results of further, longer-term evaluation of the effects of

OXM in a larger group of subjects. Moreover, OXM suffers from the same handicap as GLP-1 in that it is rapidly degraded by DPP-IV. A DPP-IV-resistant analogue would possibly allow for a reduction in the number of injections required and thus an increase in the acceptability of the treatment in a clinical setting.

Inhibition of dipeptidyl peptidase IV

In the search for an effective gut hormone-based therapy for obesity, some researchers have adopted the approach of enhancing the effectiveness of circulating endogenous gut peptides. One method of achieving this is through inhibition of the processes that inactivate the relevant hormones. As noted above, DPP-IV is a major mediator of the inactivation of the gut hormones GLP-1 and OXM and multiple inhibitors of DPP-IV have been tested in animal models. While DPP-IV inhibition improves blood glucose levels in rodent models of type 2 diabetes, the data on weight loss are more equivocal. For example, vildagliptin (Novartis, Basel, Switzerland) has the effect of decreasing blood sugar levels in insulin-resistant Zucker fatty rats [101]. However, no effect of chronic (21-day) administration on weight was noted despite increasing levels of endogenous GLP-1. Similarly, the compound FE999011 (Ferring Pharmaceuticals, Saint-Prex, Switzerland) failed to induce loss of weight in Zucker fatty rats after 21 days of treatment [102]. Administration of the compound P32/98 to VDF fa/fa Zucker rats for 3 months, however, did result in a decrease in the rate of weight gain [103]. Both lean and diet-induced obese C57BL/6J mice, administered DPP-IV inhibitor NVP DPP 728 for 8 weeks, were not significantly different to comparable controls in terms of body weight, but there was a difference in food intake, particularly towards the end of the study period [104]. It may be that the effects on weight and food intake develop over a longer time frame than the effects on glucose tolerance.

Given the large number of substrates of DPP-IV, however, longer-term safety data are needed. In particular, DPP-IV plays an important role in the immune system (in which capacity it is known as CD26), and is also responsible for the inactivation of GLP-2, another product of proglucagon cleavage that has growth-promoting effects on the intestinal mucosa [105]. In a recently-published small trial, administration of vildagliptin in combination with metformin to type 2 diabetic patients for 1 year resulted in a rise in blood pressure, necessitating additional therapy, in a small number of subjects [106]. Although the cause of this is unknown, it may be related to the cardiovascular effects of GLP-1 described above.

Amylin

Co-secreted with insulin from the β-cells of the pancreas, the peptide amylin forms the basis of Symlin (pramlintide acetate) (Amylin Pharmaceuticals Inc), a novel treatment for diabetes which has recently been granted FDA approval.

In addition to favourable effects on blood glucose, pramlintide also reduces food intake and has been shown to cause weight loss in overweight diabetic subjects [107]. Recently, Phase II clinical trials of pramlintide for the treatment of obesity have shown clinically and statistically significant weight loss in obese subjects. Further evaluation of this drug as a therapy specifically for the treatment of obesity is awaited.

Gut hormones that signal hunger

Ghrelin

Ghrelin is the endogenous ligand for the growth hormone secretagogue receptor. Principally synthesised in the stomach, it is released into the circulation in response to a fast and has been implicated in meal initiation [9]. Both human and rodent data are supportive of a role for ghrelin in the regulation of energy balance [108–110]. A possible role for ghrelin in the pathogenesis of obesity has been suggested by the observation that ghrelin levels vary inversely with body weight, and that weight loss is associated with a rise in ghrelin levels, possibly contributing to the difficulty some individuals have in maintaining weight loss once it has been achieved [111].

Antagonism of ghrelin signalling might provide an additional means of initiating (or indeed maintaining) weight loss. However, work in this field has thus far focused on the use of ghrelin in chronic conditions associated with cachexia or malnourishment [112–114].

Co-administration and combination therapy

While polypharmacy has a number of significant problems associated with it, not least issues related to patient compliance and an increased likelihood of drug interactions, there are instances in which combination therapy can be advantageous. An oft-cited example of this is the treatment of hypertension, in which modulation of a number of different systems regulating blood pressure results in a better clinical outcome than use of one class of antihypertensive alone [115]. The rationale behind combination therapy in large part arises out of the argument that use of lower doses of drugs with different mechanisms of action, compared with a higher dose of a single therapy, minimises adverse effects while maximising therapeutic effect.

The majority of obesity is multifactorial in origin and the mechanisms underlying the regulation of energy balance are complex and interdependent. In the case of appetite control by gut hormones, it may be anticipated that combination therapy with treatments based on different peptides will mirror more closely normal physiological control of food intake than would monotherapy. For example, arguably, it would be logical to combine use of drugs based on

GLP-1 and PYY_{3-36}, since both peptides are physiologically co-secreted from the same endocrine cells in the gastrointestinal tract in response to similar stimuli. Indeed, initial data are supportive of this view. Rodents administered PYY_{3-36} in combination with either exendin-4 or $GLP\text{-}1_{7-36amide}$ do indeed demonstrate additional reductions in food intake [54, 58]. Importantly, this finding also extends to humans [54].

Thus, the combination of a number of gut hormone-based therapies, and their combination in turn with other modalities of treatment and support in a multidisciplinary approach, provides a model for the management of the obese patient that is most likely to maximise weight reduction.

Conclusions

The rising prevalence of obesity and its associated burden of morbidity, mortality and healthcare costs has highlighted the need for the development of more effective weight-loss therapies than those currently available. The importance of gut hormones in the physiological regulation of energy intake offers the promise of effective additions to the physician's armamentarium of treatments. A number of therapies are currently at differing stages of development and although there are difficulties to be overcome, significant, sustainable reduction of body weight through manipulation of normal physiological pathways remains a viable possibility.

Acknowledgements
Dr O. Chaudhri is supported by a Wellcome Trust Clinical Training Fellowship. The authors would like to record their thanks to Chrystalla Orphanides and Sara Yadav for their help in the preparation of the manuscript.

References

1 National Audit Office (2001) Tackling obesity in England: a report by the Comptroller and Auditor General.
2 Mokdad AH, Mark JS, Stroup DF, Gerberding JL (2004) Actual causes of death in the United States, 2000. *JAMA* 291: 1238–1245
3 Kaplan LM (2005) Pharmacological therapies for obesity. *Gastroenterol Clin North Am* 34: 91–104
4 Thearle M, Aronne LJ (2003) Obesity and pharmacologic therapy. *Endocrinol Metab Clin North Am* 32: 1005–1024
5 Yanovski SZ, Yanovski JA (2002) Obesity. *N Engl J Med* 346: 591–602
6 Stanley S, Wynne K, McGowan B, Bloom S (2005) Hormonal regulation of food intake. *Physiol Rev* 85: 1131–1158
7 Wynne K, Stanley S, McGowan B, Bloom S (2005) Appetite control. *J Endocrinol* 184: 291–318
8 Dockray GJ (1979) Comparative biochemistry and physiology of gut hormones. *Annu Rev Physiol* 41: 83–95
9 Korbonits M, Goldstone AP, Gueorguiev M, Grossman AB (2004) Ghrelin – a hormone with multiple functions. *Front Neuroendocrinol* 25: 27–68
10 Rehfeld JF (2004) Clinical endocrinology and metabolism. Cholecystokinin. *Best Pract Res Clin Endocrinol Metab* 18: 569–586

11 Dockray GJ (2004) Clinical endocrinology and metabolism. Gastrin. *Best Pract Res Clin Endocrinol Metab* 18: 555–568
12 Geloneze B, Tambascia MA, Pilla VF, Geloneze SR, Repetto EM, Pareja JC (2003) Ghrelin: a gut-brain hormone: effect of gastric bypass surgery. *Obes Surg* 13: 17–22
13 Korner J, Bessler M, Cirilo LJ, Conwell IM, Daud A, Restuccia NL, Wardlaw SL (2005) Effects of Roux-en-Y gastric bypass surgery on fasting and postprandial concentrations of plasma ghrelin, peptide YY, and insulin. *J Clin Endocrinol Metab* 90: 359–365
14 Service GJ, Thompson GB, Service FJ, Andrews JC, Collazo-Clavell ML, Lloyd RV (2005) Hyperinsulinemic hypoglycemia with nesidioblastosis after gastric-bypass surgery. *N Engl J Med* 353: 249–254
15 Cone RD, Cowley MA, Butler AA, Fan W, Marks DL, Low MJ (2001) The arcuate nucleus as a conduit for diverse signals relevant to energy homeostasis. *Int J Obes Relat Metab Disord* 25 Suppl 5: S63–S67
16 Schwartz MW, Woods SC, Porte D Jr, Seeley RJ, Baskin DG (2000) Central nervous system control of food intake. *Nature* 404: 661–671
17 Gibbs J, Young RC, Smith GP (1973) Cholecystokinin decreases food intake in rats. *J Comp Physiol Psychol* 84: 488–495
18 Kissileff HR, Pi-Sunyer FX, Thornton J, Smith GP (1981) C-terminal octapeptide of cholecystokinin decreases food intake in man. *Am J Clin Nutr* 34: 154–160
19 McLaughlin CL, Baile CA, Buonom FC (1985) Effect of CCK antibodies on food intake and weight gain in Zucker rats. *Physiol Behav* 34: 277–282
20 Meereis-Schwanke K, Klonowski-Stumpe H, Herberg L, Niederau C (1998) Long-term effects of CCK-agonist and -antagonist on food intake and body weight in Zucker lean and obese rats. *Peptides* 19: 291–299
21 Moran TH, Katz LF, Plata-Salaman CR, Schwartz GJ (1998) Disordered food intake and obesity in rats lacking cholecystokinin A receptors. *Am J Physiol* 274: R618–R625
22 West DB, Fey D, Woods SC (1984) Cholecystokinin persistently suppresses meal size but not food intake in free-feeding rats. *Am J Physiol* 246: R776–R787
23 West DB, Greenwood MR, Sullivan AC, Prescod L, Marzullo LR, Triscari J (1987) Infusion of cholecystokinin between meals into free-feeding rats fails to prolong the intermeal interval. *Physiol Behav* 39: 111–115
24 Crawley JN, Beinfeld MC (1983) Rapid development of tolerance to the behavioural actions of cholecystokinin. *Nature* 302: 703–706
25 Fuhlendorff J, Johansen NL, Melberg SG, Thogersen H, Schwartz TW (1990) The antiparallel pancreatic polypeptide fold in the binding of neuropeptide Y to Y1 and Y2 receptors. *J Biol Chem* 265: 11706–11712
26 Michel MC, Beck-Sickinger A, Cox H, Doods HN, Herzog H, Larhammar D, Quirion R, Schwartz T, Westfall T (1998) XVI. International Union of Pharmacology recommendations for the nomenclature of neuropeptide Y, peptide YY, and pancreatic polypeptide receptors. *Pharmacol Rev* 50: 143–150
27 Adrian TE, Bloom SR, Bryant MG, Polak JM, Heitz PH, Barnes AJ (1976) Distribution and release of human pancreatic polypeptide. *Gut* 17: 940–944
28 Batterham RL, Le Roux CW, Cohen MA, Park AJ, Ellis SM, Patterson M, Frost GS, Ghatei MA, Bloom SR (2003) Pancreatic polypeptide reduces appetite and food intake in humans. *J Clin Endocrinol Metab* 88: 3989–3992
29 Asakawa A, Inui A, Yuzuriha H, Ueno N, Katsuura G, Fujimiya M, Fujino MA, Niijima A, Meguid MM, Kasuga M (2003) Characterization of the effects of pancreatic polypeptide in the regulation of energy balance. *Gastroenterology* 124: 1325–1336
30 Asakawa A, Inui A, Ueno N, Fujimiya M, Fujino MA, Kasuga M (1999) Mouse pancreatic polypeptide modulates food intake, while not influencing anxiety in mice. *Peptides* 20: 1445–1448
31 Zipf WB, O'Dorisio TM, Cataland S, Sotos J (1981) Blunted pancreatic polypeptide responses in children with obesity of Prader-Willi syndrome. *J Clin Endocrinol Metab* 52: 1264–1266
32 Fujimoto S, Inui A, Kiyota N, Seki W, Koide K, Takamiya S, Uemoto M, Nakajima Y, Baba S, Kasuga M (1997) Increased cholecystokinin and pancreatic polypeptide responses to a fat-rich meal in patients with restrictive but not bulimic anorexia nervosa. *Biol Psychiatry* 41: 1068–1070
33 Malaisse-Lagae F, Carpentier JL, Patel YC, Malaisse WJ, Orci L (1977) Pancreatic polypeptide: a possible role in the regulation of food intake in the mouse. Hypothesis. *Experientia* 33: 915–917

34 Ueno N, Inui A, Iwamoto M, Kaga T, Asakawa A, Okita M, Fujimiya M, Nakajima Y, Ohmoto Y, Ohnaka M et al. (1999) Decreased food intake and body weight in pancreatic polypeptide-overexpressing mice. *Gastroenterology* 117: 1427–1432
35 Campbell RE, Smith MS, Allen SE, Grayson BE, Ffrench-Mullen JM, Grove KL (2003) Orexin neurons express a functional pancreatic polypeptide Y4 receptor. *J Neurosci* 23: 1487–1497
36 Jorde R, Burhol PG (1984) Fasting and postprandial plasma pancreatic polypeptide (PP) levels in obesity. *Int J Obes* 8: 393–397
37 Wisen O, Bjorvell H, Cantor P, Johansson C, Theodorsson E (1992) Plasma concentrations of regulatory peptides in obesity following modified sham feeding (MSF) and a liquid test meal. *Regul Pept* 39: 43–54
38 Schmidt PT, Naslund E, Gryback P, Jacobsson H, Holst JJ, Hilsted L, Hellstrom PM (2005) A role for pancreatic polypeptide in the regulation of gastric emptying and short term metabolic control. *J Clin Endocrinol Metab* 90: 5241–5246
39 Berntson GG, Zipf WB, O'Dorisio TM, Hoffman JA, Chance RE (1993) Pancreatic polypeptide infusions reduce food intake in Prader-Willi syndrome. *Peptides* 14: 497–503
40 Adrian TE, Ferri GL, Bacarese-Hamilton AJ, Fuessl HS, Polak JM, Bloom SR (1985) Human distribution and release of a putative new gut hormone, peptide YY. *Gastroenterology* 89: 1070–1077
41 Hagan MM (2002) Peptide YY: a key mediator of orexigenic behavior. *Peptides* 23: 377–382
42 Kanatani A, Mashiko S, Murai N, Sugimoto N, Ito J, Fukuroda T, Fukami T, Morin N, MacNeil DJ, Van der Ploeg LH et al. (2000) Role of the Y1 receptor in the regulation of neuropeptide Y-mediated feeding: comparison of wild-type, Y1 receptor-deficient, and Y5 receptor-deficient mice. *Endocrinology* 141: 1011–1016
43 Batterham RL, Cowley MA, Small CJ, Herzog H, Cohen MA, Dakin CL, Wren AM, Brynes AE, Low MJ, Ghatei MA et al. (2002) Gut hormone PYY(3–36) physiologically inhibits food intake. *Nature* 418: 650–654
44 Abbott CR, Small CJ, Kennedy AR, Neary NM, Sajedi A, Ghatei MA, Bloom SR (2005) Blockade of the neuropeptide Y Y2 receptor with the specific antagonist BIIE0246 attenuates the effect of endogenous and exogenous peptide YY(3–36) on food intake. *Brain Res* 1043: 139–144
45 Scott V, Kimura N, Stark JA, Luckman SM (2005) Intravenous peptide YY3–36 and Y2 receptor antagonism in the rat: effects on feeding behaviour. *J Neuroendocrinol* 17: 452–457
46 Nonaka N, Shioda S, Niehoff ML, Banks WA (2003) Characterization of blood-brain barrier permeability to PYY3–36 in the mouse. *J Pharmacol Exp Ther* 306: 948–953
47 Koda S, Date Y, Murakami N, Shimbara T, Hanada T, Toshinai K, Niijima A, Furuya M, Inomata N, Osuye K et al. (2005) The role of the vagal nerve in peripheral PYY3–36-induced feeding reduction in rats. *Endocrinology* 146: 2369–2375
48 Abbott CR, Monteiro M, Small CJ, Sajedi A, Smith KL, Parkinson JR, Ghatei MA, Bloom SR (2005) The inhibitory effects of peripheral administration of peptide YY(3–36) and glucagon-like peptide-1 on food intake are attenuated by ablation of the vagal-brainstem-hypothalamic pathway. *Brain Res* 1044: 127–131
49 Halatchev IG, Cone RD (2005) Peripheral administration of PYY(3–36) produces conditioned taste aversion in mice. *Cell Metab* 1: 159–168
50 Cox JE, Randich A (2004) Enhancement of feeding suppression by PYY(3–36) in rats with area postrema ablations. *Peptides* 25: 985–989
51 Batterham RL, Cohen MA, Ellis SM, Le Roux CW, Withers DJ, Frost GS, Ghatei MA, Bloom SR (2003) Inhibition of food intake in obese subjects by peptide YY3–36. *N Engl J Med* 349: 941–948
52 le Roux CW, Batterham RL, Aylwin SJ, Patterson M, Borg CM, Wynne KJ, Kent A, Vincent RP, Gardiner J, Ghatei MA et al. (2005) Attenuated peptide YY release in obese subjects is associated with reduced satiety. *Endocrinology* 147: 3–8
53 Boggiano MM, Chandler PC, Oswald KD, Rodgers RJ, Blundell JE, Ishii Y, Beattie AH, Holch P, Allison DB, Schindler M et al. (2005) PYY3–36 as an anti-obesity drug target. *Obes Rev* 6: 307–322
54 Neary NM, Small CJ, Druce MR, Park AJ, Ellis SM, Semjonous NM, Dakin CL, Filipsson K, Wang F, Kent AS et al. (2005) Peptide YY3–36 and glucagon-like peptide-17–36 inhibit food intake additively. *Endocrinology* 146: 5120–5127
55 Nordheim U, Hofbauer KG (2004) Stimulation of NPY Y2 receptors by PYY3–36 reveals divergent cardiovascular effects of endogenous NPY in rats on different dietary regimens. *Am J Physiol Regul Integr Comp Physiol* 286: R138–R142
56 Chelikani PK, Haver AC, Reidelberger RD (2005) Intravenous infusion of peptide YY(3–36)

potently inhibits food intake in rats. *Endocrinology* 146: 879–888
57 Challis BG, Pinnock SB, Coll AP, Carter RN, Dickson SL, O'Rahilly S (2003) Acute effects of PYY3–36 on food intake and hypothalamic neuropeptide expression in the mouse. *Biochem Biophys Res Commun* 311: 915–919
58 Talsania T, Anini Y, Siu S, Drucker DJ, Brubaker PL (2005) Peripheral exendin-4 and peptide YY3–36 synergistically reduce food intake through different mechanisms in mice. *Endocrinology* 146: 3748–3756
59 Halatchev IG, Ellacott KL, Fan W, Cone RD (2004) Peptide YY3–36 inhibits food intake in mice through a melanocortin-4 receptor-independent mechanism. *Endocrinology* 145: 2585–2590
60 Shechter Y, Tsubery H, Mironchik M, Rubinstein M, Fridkin M (2005) Reversible PEGylation of peptide YY3–36 prolongs its inhibition of food intake in mice. *FEBS Lett* 579: 2439–2444
61 Riediger T, Bothe C, Becskei C, Lutz TA (2004) Peptide YY directly inhibits ghrelin-activated neurons of the arcuate nucleus and reverses fasting-induced c-Fos expression. *Neuroendocrinology* 79: 317–326
62 Challis BG, Coll AP, Yeo GS, Pinnock SB, Dickson SL, Thresher RR, Dixon J, Zahn D, Rochford JJ, White A et al. (2004) Mice lacking pro-opiomelanocortin are sensitive to high-fat feeding but respond normally to the acute anorectic effects of peptide-YY(3–36). *Proc Natl Acad Sci USA* 101: 4695–4700
63 Adams SH, Won WB, Schonhoff SE, Leiter AB, Paterniti JR Jr, (2004) Effects of peptide YY[3–36] on short-term food intake in mice are not affected by prevailing plasma ghrelin levels. *Endocrinology* 145: 4967–4975
64 Pittner RA, Moore CX, Bhavsar SP, Gedulin BR, Smith PA, Jodka CM, Parkes DG, Paterniti JR, Srivastava VP, Young AA (2004) Effects of PYY[3–36] in rodent models of diabetes and obesity. *Int J Obes Relat Metab Disord* 28: 963–971
65 Martin NM, Small CJ, Sajedi A, Patterson M, Ghatei MA, Bloom SR (2004) Pre-obese and obese agouti mice are sensitive to the anorectic effects of peptide YY(3–36) but resistant to ghrelin. *Int J Obes Relat Metab Disord* 28: 886–893
66 Abbott CR, Small CJ, Sajedi A, Smith KL, Parkinson JR, Broadhead LL, Ghatei MA, Bloom SR (2005) The importance of acclimatisation and habituation to experimental conditions when investigating the anorectic effects of gastrointestinal hormones in the rat. *Int J Obes (Lond)* 30: 288–292
67 Kas MJ, Bruijnzeel AW, Haanstra JR, Wiegant VM, Adan RA (2005) Differential regulation of agouti-related protein and neuropeptide Y in hypothalamic neurons following a stressful event. *J Mol Endocrinol* 35: 159–164
68 Ghamari-Langroudi M, Colmers WF, Cone RD (2005) PYY3–36 inhibits the action potential firing activity of POMC neurons of arcuate nucleus through postsynaptic Y2 receptors. *Cell Metab* 2: 191–199
69 Acuna-Goycolea C, van den Pol AN (2005) Peptide YY(3–36) inhibits both anorexigenic proopiomelanocortin and orexigenic neuropeptide Y neurons: implications for hypothalamic regulation of energy homeostasis. *J Neurosci* 25: 10510–10519
70 Moran TH, Smedh U, Kinzig KP, Scott KA, Knipp S, Ladenheim EE (2004) Peptide YY (3–36) inhibits gastric emptying and produces acute reductions in food intake in rhesus monkeys. *Am J Physiol Regul Integr Comp Physiol* 288: R384–388
71 Sileno AP, Brandt GC, Spann BM, Quay SC (2005) Lower mean weight after 14 days intravenous administration peptide YY(3–36) (PYY(3–36)) in rabbits. *Int J Obes (Lond)* 30: 68–72
72 Koegler FH, Enriori PJ, Billes SK, Takahashi DL, Martin MS, Clark RL, Evans AE, Grove KL, Cameron L, Cowley MA (2005) Peptide YY(3–36) inhibits morning, but not evening, food intake and decreases body weight in rhesus macaques. *Diabetes* 54: 3198–3204
73 Brandt G, Park A, Wynne K, Sileno A, Jazrawi R, Woods A, Quay S, Bloom S (2004) Nasal peptide YY3–36: Phase 1 dose ranging and safety studies in healthy human subjects. *86th Annual Meeting of the Endocrine Society (ENDO 2004)*, New Orleans, LA
74 Kieffer TJ, Habener JF (1999) The glucagon-like peptides. *Endocr Rev* 20: 876–913
75 Eissele R, Goke R, Willemer S, Harthus HP, Vermeer H, Arnold R, Goke B (1992) Glucagon-like peptide-1 cells in the gastrointestinal tract and pancreas of rat, pig and man. *Eur J Clin Invest* 22: 283–291
76 Orskov C, Rabenhoj L, Wettergren A, Kofod H, Holst JJ (1994) Tissue and plasma concentrations of amidated and glycine-extended glucagon-like peptide I in humans. *Diabetes* 43: 535–539
77 Holst JJ (2005) Glucagon-like peptide-1: physiology and therapeutic potential. *Curr Opin Endocrinol Diabetes* 12: 56–62

78 Wei Y, Mojsov S (1995) Tissue-specific expression of the human receptor for glucagon-like peptide-I: brain, heart and pancreatic forms have the same deduced amino acid sequences. *FEBS Lett* 358: 219–224
79 Shughrue PJ, Lane MV, Merchenthaler I (1996) Glucagon-like peptide-1 receptor (GLP1-R) mRNA in the rat hypothalamus. *Endocrinology* 137: 5159–5162
80 Turton MD, O'Shea D, Gunn I, Beak SA, Edwards CM, Meeran K, Choi SJ, Taylor GM, Heath MM, Lambert PD et al. (1996) A role for glucagon-like peptide-1 in the central regulation of feeding. *Nature* 379: 69–72
81 Meeran K, O'Shea D, Edwards CM, Turton MD, Heath MM, Gunn I, Abusnana S, Rossi M, Small CJ, Goldstone AP et al. (1999) Repeated intracerebroventricular administration of glucagon-like peptide-1-(7–36) amide or exendin-(9–39) alters body weight in the rat. *Endocrinology* 140: 244–250
82 Scrocchi LA, Brown TJ, MaClusky N, Brubaker PL, Auerbach AB, Joyner AL, Drucker DJ (1996) Glucose intolerance but normal satiety in mice with a null mutation in the glucagon-like peptide 1 receptor gene. *Nat Med* 2: 1254–1258
83 Kinzig KP, D'Alessio DA, Seeley RJ (2002) The diverse roles of specific GLP-1 receptors in the control of food intake and the response to visceral illness. *J Neurosci* 22: 10470–10476
84 Verdich C, Flint A, Gutzwiller JP, Naslund E, Beglinger C, Hellstrom PM, Long SJ, Morgan LM, Holst JJ, Astrup A (2001) A meta-analysis of the effect of glucagon-like peptide-1 (7–36) amide on *ad libitum* energy intake in humans. *J Clin Endocrinol Metab* 86: 4382–4389
85 Verdich C, Toubro S, Buemann B, Lysgard MJ, Juul HJ, Astrup A (2001) The role of postprandial releases of insulin and incretin hormones in meal-induced satiety – effect of obesity and weight reduction. *Int J Obes Relat Metab Disord* 25: 1206–1214
86 Feinle C, Chapman IM, Wishart J, Horowitz M (2002) Plasma glucagon-like peptide-1 (GLP-1) responses to duodenal fat and glucose infusions in lean and obese men. *Peptides* 23: 1491–1495
87 Zander M, Madsbad S, Madsen JL, Holst JJ (2002) Effect of 6-week course of glucagon-like peptide 1 on glycaemic control, insulin sensitivity, and beta-cell function in type 2 diabetes: a parallel-group study. *Lancet* 359: 824–830
88 Naslund E, King N, Mansten S, Adner N, Holst JJ, Gutniak M, Hellstrom PM (2004) Prandial subcutaneous injections of glucagon-like peptide-1 cause weight loss in obese human subjects. *Br J Nutr* 91: 439–446
89 Nauck MA, Meier JJ (2005) Glucagon-like peptide 1 and its derivatives in the treatment of diabetes. *Regul Pept* 128: 135–148
90 Purnell JQ, Weyer C (2003) Weight effect of current and experimental drugs for diabetes mellitus: from promotion to alleviation of obesity. *Treat Endocrinol* 2: 33–47
91 Larsen PJ, Fledelius C, Knudsen LB, Tang-Christensen M (2001) Systemic administration of the long-acting GLP-1 derivative NN2211 induces lasting and reversible weight loss in both normal and obese rats. *Diabetes* 50: 2530–2539
92 Fineman MS, Shen LZ, Taylor K, Kim DD, Baron AD (2004) Effectiveness of progressive dose-escalation of exenatide (exendin-4) in reducing dose-limiting side effects in subjects with type 2 diabetes. *Diabetes Metab Res Rev* 20: 411–417
93 Yamamoto H, Kishi T, Lee CE, Choi BJ, Fang H, Hollenberg AN, Drucker DJ, Elmquist JK (2003) Glucagon-like peptide-1-responsive catecholamine neurons in the area postrema link peripheral glucagon-like peptide-1 with central autonomic control sites. *J Neurosci* 23: 2939–2946
94 Edwards CM, Edwards AV, Bloom R (1997) Cardiovascular and pancreatic endocrine responses to glucagon-like peptide-1(7–36) amide in the conscious calf. *Exp Physiol* 82: 709–716
95 Dakin CL, Gunn I, Small CJ, Edwards CM, Hay DL, Smith DM, Ghatei MA, Bloom SR (2001) Oxyntomodulin inhibits food intake in the rat. *Endocrinology* 142: 4244–4250
96 Dakin CL, Small CJ, Batterham RL, Neary NM, Cohen MA, Patterson M, Ghatei MA, Bloom SR (2004) Peripheral oxyntomodulin reduces food intake and body weight gain in rats. *Endocrinology* 145: 2687–2695
97 Baggio LL, Huang Q, Brown TJ, Drucker DJ (2004) Oxyntomodulin and glucagon-like peptide-1 differentially regulate murine food intake and energy expenditure. *Gastroenterology* 127: 546–558
98 Cohen MA, Ellis SM, Le Roux CW, Batterham RL, Park A, Patterson M, Frost GS, Ghatei MA, Bloom SR (2003) Oxyntomodulin suppresses appetite and reduces food intake in humans. *J Clin Endocrinol Metab* 88: 4696–4701
99 Wynne K, Park AJ, Small CJ, Patterson M, Ellis SM, Murphy KG, Wren AM, Frost GS, Meeran

K, Ghatei MA et al. (2005) Subcutaneous oxyntomodulin reduces body weight in overweight and obese subjects: a double-blind, randomized, controlled trial. *Diabetes* 54: 2390–2395
100 Dakin CL, Small CJ, Park AJ, Seth A, Ghatei MA, Bloom SR (2002) Repeated ICV administration of oxyntomodulin causes a greater reduction in body weight gain than in pair-fed rats. *Am J Physiol Endocrinol Metab* 283: E1173–E1177
101 Burkey BF, Li X, Bolognese L, Balkan B, Mone M, Russell M, Hughes TE, Wang PR (2005) Acute and chronic effects of the incretin enhancer vildagliptin in insulin-resistant rats. *J Pharmacol Exp Ther* 315: 688–695
102 Sudre B, Broqua P, White RB, Ashworth D, Evans DM, Haigh R, Junien JL, Aubert ML (2002) Chronic inhibition of circulating dipeptidyl peptidase IV by FE 999011 delays the occurrence of diabetes in male zucker diabetic fatty rats. *Diabetes* 51: 1461–1469
103 Pospisilik JA, Stafford SG, Demuth HU, McIntosh CH, Pederson RA (2002) Long-term treatment with dipeptidyl peptidase IV inhibitor improves hepatic and peripheral insulin sensitivity in the VDF Zucker rat: a euglycemic-hyperinsulinemic clamp study. *Diabetes* 51: 2677–2683
104 Reimer MK, Holst JJ, Ahren B (2002) Long-term inhibition of dipeptidyl peptidase IV improves glucose tolerance and preserves islet function in mice. *Eur J Endocrinol* 146: 717–727
105 Drucker DJ (2001) Glucagon-like peptide 2. *J Clin Endocrinol Metab* 86: 1759–1764
106 Ahren B, Gomis R, Standl E, Mills D, Schweizer A (2004) Twelve- and 52-week efficacy of the dipeptidyl peptidase IV inhibitor LAF237 in metformin-treated patients with type 2 diabetes. *Diabetes Care* 27: 2874–2880
107 Hollander P, Maggs DG, Ruggles JA, Fineman M, Shen L, Kolterman OG, Weyer C (2004) Effect of pramlintide on weight in overweight and obese insulin-treated type 2 diabetes patients. *Obes Res* 12: 661–668
108 Wren AM, Small CJ, Abbott CR, Dhillo WS, Seal LJ, Cohen MA, Batterham RL, Taheri S, Stanley SA, Ghatei MA et al. (2001) Ghrelin causes hyperphagia and obesity in rats. *Diabetes* 50: 2540–2547
109 Wren AM, Seal LJ, Cohen MA, Brynes AE, Frost GS, Murphy KG, Dhillo WS, Ghatei MA, Bloom SR (2001) Ghrelin enhances appetite and increases food intake in humans. *J Clin Endocrinol Metab* 86: 5992–5995
110 Tschop M, Smiley DL, Heiman ML (2000) Ghrelin induces adiposity in rodents. *Nature* 407: 908–913
111 Hansen TK, Dall R, Hosoda H, Kojima M, Kangawa K, Christiansen JS, Jorgensen JO (2002) Weight loss increases circulating levels of ghrelin in human obesity. *Clin Endocrinol (Oxf)* 56: 203–206
112 Neary NM, Small CJ, Wren AM, Lee JL, Druce MR, Palmieri C, Frost GS, Ghatei MA, Coombes RC, Bloom SR (2004) Ghrelin increases energy intake in cancer patients with impaired appetite: acute, randomized, placebo-controlled trial. *J Clin Endocrinol Metab* 89: 2832–2836
113 Nagaya N, Itoh T, Murakami S, Oya H, Uematsu M, Miyatake K, Kangawa K (2005) Treatment of cachexia with ghrelin in patients with COPD. *Chest* 128: 1187–1193
114 Wynne K, Giannitsopoulou K, Small CJ, Patterson M, Frost G, Ghatei MA, Brown EA, Bloom SR, Choi P (2005) Subcutaneous ghrelin enhances acute food intake in malnourished patients who receive maintenance peritoneal dialysis: a randomized, placebo-controlled trial. *J Am Soc Nephrol* 16: 2111–2118
115 Ogihara T, Matsuzaki M, Matsuoka H, Shimamoto K, Shimada K, Rakugi H, Umemoto S, Kamiya A, Suzuki N, Kumagai H et al. (2005) The combination therapy of hypertension to prevent cardiovascular events (COPE) trial: rationale and design. *Hypertens Res* 28: 331–338

Pharmacotherapy of Obesity
Edited by J.P.H. Wilding

Influencing energy expenditure and substrate utilisation

John C. Clapham[1] and Jonathan R. Arch[2]

[1] *CVG I Bioscience, Astra Zeneca, Mereside, Alderley Park, Macclesfield SK10 4TG, UK*
[2] *Clore Laboratory, University of Buckingham, Buckingham, Bucks MK18 1EG, UK*

Introduction

Most of the anti-obesity drugs used over the past 50 years, and one (sibutramine) of the two globally licensed for long-term use today have acted centrally. With the exception of sibutramine (and the possible further approval of rimonabant), however, all of these have been withdrawn due to concerns over toxicity or abuse potential, or their use is greatly restricted. This is one reason why pharmaceutical companies are seeking drugs that act peripherally to influence energy expenditure, substrate utilisation or both. Fortunately, studies on genetically modified mice are revealing many new targets that fall into these categories [1].

It should not be assumed that centrally acting drugs only influence food intake. There is ample evidence that many anorectic drugs also increase energy expenditure in rodents [2], and possibly also in humans [3]. There is also good evidence that alteration of substrate availability in the hypothalamus can affect both energy intake and energy expenditure [4]. However, with the exception of some hormone mimetics (such as leptin mimetics), the focus of this chapter is on peripheral targets.

Rationale for thermogenic drugs

There are aetiological arguments for thermogenic drugs being a rational approach to the treatment of obesity (Tab. 1). Firstly, reduced energy expenditure is partly responsible for the obesity epidemic. Secondly, susceptibility to obesity at any point in time is inversely related to daily energy expenditure [5]. Variation in daily energy expenditure may be primarily due to variation in locomotor activity [6], which is not obviously amenable to pharmacotherapy. There is also evidence, however, that low resting metabolic rate – seen in both pre-obese and post-obese subjects – predicts the development of obesity [7, 8], despite the fact that once they become obese, humans tend to have an elevated

Table 1. Reasons for seeking peripherally acting thermogenic drugs

Pragmatic	• Risks associated with central actions • Many peripheral targets identified – especially using genetically modified mice
Aetiological	• Decreasing energy expenditure is one cause of the rise in obesity • Susceptibility to obesity is associated with low energy expenditure • Susceptibility to obesity is associated with poor ability to oxidise fat
Therapeutic	• Stimulants of fat oxidation cause loss of fat only • Stimulation of fat oxidation is especially beneficial for insulin sensitivity and hypertriglyceridaemia • Counter-regulatory forces *may* be less than with anorectic agents so that weight loss is more prolonged

metabolic rate and so must consume more energy than lean subjects to maintain their obesity. There are examples of genetically modified mice in which hyperphagia develops after obesity [1].

A third argument, specifically for thermogenic drugs that stimulate fat oxidation, is that fat oxidation is defective in people who are susceptible to obesity, both when they are obese and after they have slimmed. This defect is apparent both at the whole body level [9] and in muscle biopsy samples [10]. The muscle findings may reflect differences in the proportions of oxidative and glycolytic fibres in those who are resistant and susceptible to obesity. For example, following bariatric surgery (e.g., gastric banding) weight loss was greatest in patients with the highest proportion of type I oxidative fibres in the rectus abdominus muscle [11].

Therapeutic benefits

In addition to pragmatic and aetiological arguments, there are therapeutic arguments for seeking thermogenic drugs – at least those that promote fat oxidation. Sympathomimetic agents, such as β_3-adrenoceptor agonists, depend on fat oxidation for thermogenesis, and, provided they do not decrease food intake, all the weight loss that they produce is fat [12]. Weight loss in response to a moderately reduced energy intake, by contrast, typically includes about 15–25% of fat-free mass [13]. In addition, β_3-adrenoceptor agonists improve insulin sensitivity beyond what would be expected from reduced body weight [12]. This may be because there is a rapid fall in the concentration of lipid metabolites, such as diacylglycerol, that are known to inhibit insulin signalling [12, 14]. By contrast, the triglyceride stores that cause adiposity are so large that it must take longer for increased fat oxidation to deplete them. It is possible that these benefits of sympathomimetic drugs will extend to other agents that promote fat oxidation.

Thermogenic drugs that mimic the sympathetic nervous system or depend on the capacity for fat oxidation may not cause as much weight loss as anorectic drugs in the short term. Past and present anorectic drugs initially produce weight loss of about 0.23 kg per week [15]. If 15% of this weight loss is lean tissue, which consists of 75% water and contains 1 kcal per gram, and 85% is fat, which contains 9 kcal per gram, the loss of energy per day is 256 kcal. This is roughly a 10% alteration in energy balance for an obese subject. The sympathetic nervous system, which depends on fat oxidation, seems to increase energy expenditure by 30% at most in humans [16]. A drug acting by a similar mechanism would do well to use a third of this capacity. Moreover, if the weight loss is all fat rather than 85% fat and 15% lean, weight loss is reduced by 13% for the same effect on energy balance.

On the other hand, it is possible that weight loss in response to a thermogenic drug would last longer than the 6 months generally seen with anorectic drugs. In rodents at least, the effect of sympathomimetic agents increases in the early stages of treatment provided animals are obese and can therefore provide fuel for thermogenesis [12]. Obese animals do not normally compensate by eating more, just as moderate exercise in obese humans does not in the short term increase food intake [17]. By contrast, most anorectic agents reduce food intake for only a few days in lean or obese rodents. This does not mean that anorectic agents are no longer influencing food intake: body weight stays below the control level and if the anorectic agent is withdrawn there is a rebound hyperphagia. It seems, therefore, that the effect of the anorectic agent meets opposition from counter-regulatory forces. Where such forces (e.g., a low plasma leptin level) are due to loss of fat, they will oppose both thermogenic and anorectic agents. However, signals that indicate that the gut is empty, or that an individual is hedonically or socially deprived, will only oppose anorectic drugs.

It is important to note that if a thermogenic drug increases energy expenditure by 10% or even 20%, it is unlikely to produce discomfort. In one study 'sitting and fidgeting' increased energy expenditure by 54%, and walking at only 1.6 km/h increased it by 154% compared to lying down [6]. Of course, this means that those who fidget or walk for much of the day are unlikely to be obese.

Biochemistry and endocrinology of thermogenesis

The great majority of the chemical energy in macronutrients is funnelled through NADH or (in the mitochondrion) FADH. Most of the energy carried by these coenzymes is used to synthesise ATP, mainly by oxidative phosphorylation in the mitochondrion. A small amount of energy is funnelled through NADPH, which is used in various anabolic processes. Together with NADH, NADPH can also be consumed by non-mitochondrial oxidases that do not conserve chemical energy by producing ATP (Fig. 1).

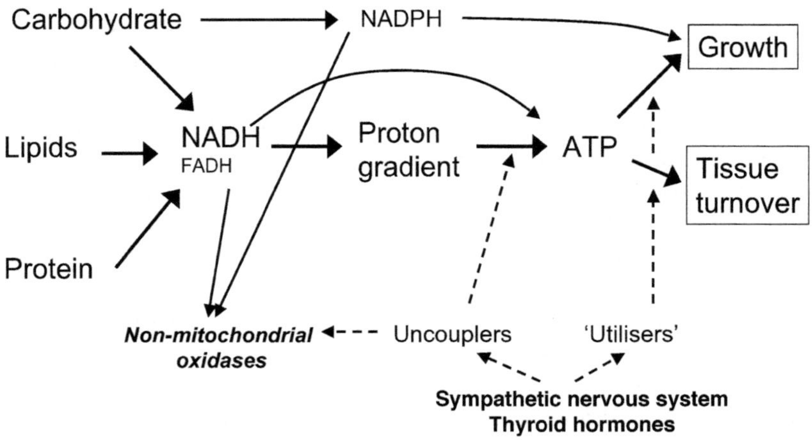

Figure 1. Production and utilisation of reduced adenine dinucleotides and ATP. Hormones and drugs that increase energy expenditure must either increase the utilisation of ATP or the oxidation of reduced coenzymes by pathways that are not coupled to ATP synthesis.

Thermogenic agents must either increase the utilisation of ATP or the oxidation of the reduced coenzymes by pathways that are not coupled to ATP synthesis, for example by uncoupling oxidative phosphorylation. Excluding growth, in which some of the energy in ATP is retained, ATP utilisation may involve ion transport, simple substrate cycles or more complex cycles. An example of a simple substrate cycle is phosphorylation of fructose-6-phosphate to fructose-1,6,-bisphosphate and hydrolysis of the bisphosphate back to the monophosphate. More complex cycles include triglyceride and protein turnover, and the Cori cycle, in which glucose and lactate are transferred between muscle and liver.

In the resting state, about 90% of oxygen consumption takes place in the mitochondrion and 80% of this is coupled to ATP synthesis. About 30% of the ATP is used for Na^+/K^+ and Ca^{2+} pumps, 30% for protein synthesis, and 20% for gluconeogenesis, ureagenesis and glycogen and triglyceride turnover [18].

Resting metabolic rate in humans can be increased by about 30% by the sympathetic nervous system [16] and by about 15% by growth hormone [19]. Thyroid hormones can increase metabolic rate by about 15% and hypothyroidism reduces it by 30% [20]. Glucocorticoids suppress energy expenditure in rodents, but raise it in humans [21]. The biochemical mechanisms by which these hormones affect metabolic rate are unclear. Uncoupling of oxidative phosphorylation may predominate in rodents, but in humans ATP consuming mechanism probably assume greater importance [22].

Hormone mimetics

Thyroid hormones

Thyroid hormones were first used to treat obesity in the 1890 s but they should only be used today in hypothyroid subjects. They have adverse cardiac effects and in some studies cause excessive loss of lean tissue. Selective stimulation of thyroid hormone receptor-β may avoid the cardiac effects of thyroid hormones [23].

Sympathetic nervous system

Sibutramine increases sympathetic activity [3]. This may contribute to its therapeutic benefits (see Tab. 1), but increased sympathetic outflow is not just to tissues that are important for thermogenesis: heart rate and blood pressure are also raised. Indeed, it is advisable to be cautious when any novel, centrally acting agent raises energy expenditure, because this may be due to a generalised increase in sympathetic activity.

β_3-Adrenoceptor agonists avoid the cardiovascular side-effects associated with β_1- and β_2-adrenoceptor stimulation. They work well in rodents, but unfortunately, the human β_3-adrenoceptor differs from the rodent receptor and those compounds that selectively activate the human receptor have mostly displayed poor oral bioavailability or have been rapidly excreted. Moreover, β_3-adrenoceptor agonists have less effect on energy expenditure in humans than in rodents. Weight loss following 28 days treatment with the β-adrenoceptor agonist L-796568 was correlated with the blood level of the compound [24], but the results were not sufficiently exciting to merit a longer trial. A longer (12 week) trial with the non-selective sympathomimetic agent ephedrine did produce significant weight loss [25].

Lipid-mobilising factor/Zn α_2 glycoprotein is produced by tumours and adipose tissue and has similar effects on energy balance and body composition to β_3-adrenoceptor agonists. Indeed, it has been claimed to stimulate β_3-adrenoceptors [26]. As an injectable therapy, perhaps it could be successful where β_3-adrenoceptor agonists have so far failed.

Growth hormone

AOD 9401 is a synthetic analogue of growth hormone that does not affect blood glucose or growth. Over 12 weeks at a dose level of 1 mg per day, weight loss was 2 kg relative to placebo treatment. Higher doses were ineffective, however, and the obvious suggestion that that was because fat loss was countered by gain of lean tissue seems to have been ruled out [27]. A further study is being conducted at the lower dose levels.

Leptin

Leptin is released mainly from adipocytes and signals to the brain whether fat stores are adequate. In rodents, leptin both decreases food intake and increases energy expenditure, the latter effect being mainly due to increased sympathetic activity. Presumably, leptin might have similar effects in humans if they are thin enough to have low leptin levels, but marked effects of leptin on energy balance have been demonstrated only in rare obese individuals that totally lack leptin. Leptin and modified leptins have not proved of benefit in most obese subjects [28]. Low molecular weight leptin mimetics that bypass the transporter that takes leptin into the brain may have greater potential [29].

Ciliary neurotrophic factor shares some signalling mechanisms with leptin. Axokine is a genetically engineered variant of human CNTF. Like leptin, it both decreases food intake and increases energy expenditure in rodents, but it is effective in rodent models of obesity that are resistant to leptin [30]. Axokine performed well in Phase II clinical trials but less well in longer Phase III trials, possibly due to the generation of neutralising antibodies.

Adiponectin

Adiponectin is a protein that is secreted by adipose tissue, especially by small adipocytes. It reduces weight gain and adipose tissue mass by stimulating fatty acid oxidation in muscle and liver. It seems to play an important role in the insulin sensitising effects of thiazolidinedione anti-diabetic drugs (though these cause weight gain through other mechanisms). The discovery of two receptors for adiponectin provides possible new targets, though whether these receptors are amenable to small molecular weight activators remains to be seen [31].

Glucocorticoids

The glucocorticoid system is usually seen as offering targets for anti-diabetic rather than anti-obesity drugs. It clearly has a profound effect on energy balance, however, as evidenced by visceral obesity in Cushing's syndrome and the ability of adrenalectomy to prevent most forms of rodent obesity. 11-β-hydroxysteroid dehydrogenase type 1 catalyses the conversion of inactive glucocorticoids (cortisone in humans; 11-dehydrocorticosterone in rodents) to active forms (cortisol and corticosterone). Inhibitors may reduce glucocorticoid receptor stimulation within tissues, but have little effect on plasma glucocorticoid concentrations or on feedback inhibition of the hypothalamic-pituitary-adrenal axis. Mice that lack 11-β-hydroxysteroid dehydrogenase type 1 have raised energy expenditure and are protected from diet-induced obesity [32]. Inhibitors of the enzyme increased energy expenditure in rodents [33]. Whether these effects are peripherally or centrally mediated, and whether they will translate to humans remains to be established.

Intracellular targets for thermogenic drugs

Uncoupling agents

Dinitrophenol, which uncouples the oxidation of NADH and FADH from ATP synthesis in the mitochondrion, was used as an anti-obesity agent in the 1930s. Weight loss was as much as 3 kg per week, but side effects included sweating and, more seriously, hypoxia at higher doses [34].

Perhaps lower doses of dinitrophenol or a similar coupling agent might retain adequate efficacy and also be safe. In recent years, however, there has been greater interest in the uncoupling proteins (UCPs), which appear to offer safer approaches. UCP-1 clearly plays a role in the regulation of energy expenditure and body weight in rodents, but is less important in humans. UCP-2 and UCP-3 can uncouple oxidative phosphorylation, and when UCP-3 was overexpressed in skeletal muscle it protected mice from diet-induced obesity [35]. Whether either UCP-2 and UCP-3 are normally involved in the regulation of energy expenditure is questionable, however [36]. In any event, there has been little progress in identifying activators of these molecules.

AMP-activated protein kinase

AMP-activated protein kinase (AMPK) is currently a popular target. It is activated by various upstream kinases, including LKB1, a tumour suppressor protein kinase. AMPK appears to play a central role in responding to increased energy requirements by activating enzymes that regulate catabolic, energy producing pathways, and inhibiting anabolic, energy requiring pathways. Its targets include acetyl-CoA carboxylase-2 (page 109), which is inhibited by phosphorylation, and hexokinase II and UCP-3, which are transcriptionally up-regulated [37]. It may also play a role in mitochondrial biogenesis in response to endurance training.

Recent work implicates AMPK in the mechanism of action of the anti-diabetic agent metformin, which has a small anti-obesity effect, and the adipokines leptin and adiponectin. These compounds do not activate AMPK directly. By contrast, 5-aminoimidazole-4-carboxamide-1-β-D-ribofuranoside (AICAR) is phosphorylated within cells to a mimic of AMP that is a potent activator of AMPK. AICAR may activate other enzymes as well, but it is notable that in rodents it produces similar metabolic effects to metformin [38]. However, AICAR, unlike metformin, stimulated glucose output from the liver [39]. If all AMPK activators do this, they will probably be unsuitable as anti-obesity agents.

Mitochondrial biogenesis

Various proteins might be targeted to promote mitochondrial biogenesis or the proportion of type 1 fibres in skeletal muscle. One of these is peroxisome proliferator-activated receptor γ coactivator-1α (PGC-1α). This is a transcriptional activator that interacts with a number of nuclear hormone receptors. It is found mainly in type I rather than type II skeletal muscle fibres, and its expression is increased by cold and exercise. Overexpression of PGC-1 in type II fibres resulted in their conversion to a more type I-like phenotype [40].

PGC-1α itself may not be amenable to small molecular weight ligands because it has no natural small ligands. However, both PGC-1α and the related PGC-1β enhance the activity of oestrogen-related receptor α (ERRα), which has a similar tissue distribution to PGC-1α and promotes fatty acid oxidation [41]. An ERR inverse agonist or siRNA directed against ERRα blocked the induction of oxidative phosphorylation genes by PGC-1 [42], supporting the case for ERRα agonists, although surprisingly, mice that lack ERRα are resistant to high fat diet-induced obesity [43].

PGC-1α coactivates peroxisome proliferator-activated receptor δ (PPARδ), which may be another mediator of its effects on fatty acid oxidation. Overexpression of PPARδ in skeletal muscle increases the proportion of type I fibres and protects mice from diet-induced obesity. A PPARδ agonist also gave protection [44]. A number of companies are attempting to develop PPARδ agonists, possibly incorporating activity at other PPARs to improve activity against diabetes and dyslipidaemia. Two notes of caution must be mentioned, however. First regulatory authorities are concerned about the potential carcinogenicity of novel PPAR agonists of all types. Secondly, bezafibrate is claimed to be an agonist of both PPARα and PPARδ [45], but despite being used in humans for many years, it has not been reported to have an anti-obesity effect. It is possible however, that bezafibrate is used clinically at a dose that allows it to act as a PPARα but not as a PPARδ agonist.

PGC-1α and PPARδ promote mitochondrial biogenesis in adipose tissue as well as skeletal muscle, in effect converting white into brown adipose tissue, whose main role is lipid oxidation. Overexpression of PPARδ in adipose tissue alone is sufficient to give protection from diet-induced obesity. Various other transcription factors and their coactivators and corepressors also affect mitochondrial biogenesis in skeletal muscle and adipose tissue, and may offer, or point to, targets for anti-obesity drugs [46].

Inhibitors of glutamine:fructose-6-phosphate amidotransferase (GFAT) might also promote mitochondrial biogenesis in skeletal muscle. GFAT regulates the rate of hexosamine synthesis. The terminal metabolite of this pathway is UDP-N-acetyl glucosamine, which glycosylates various proteins involved in nutrient sensing. Infusion of glucosamine decreased the expression of genes involve in oxidative phosphorylation in rat skeletal muscle, and whole body energy expenditure was reduced. Moreover, overexpression of

GFAT in liver resulted in obesity and other features of the metabolic syndrome [47].

Inhibition of lipid synthesis

Paradoxically, inhibitors of fatty acid and triglycerol synthesis (Fig. 2) usually alter energy balance in rodents by promoting fat oxidation, though some also inhibit food intake. (–)-Hydroxycitric acid is an example of a compound that appears to have both actions. It is an inhibitor of ATP citrate lyase, a key enzyme of fatty acid synthesis, and is marketed as a natural supplement for weight management [48]. Like some other targets in lipid synthetic pathways, it is unclear whether peripheral (adipose tissue or liver) or hypothalamic ATP citrate lyase plays the primary role in influencing energy balance [49].

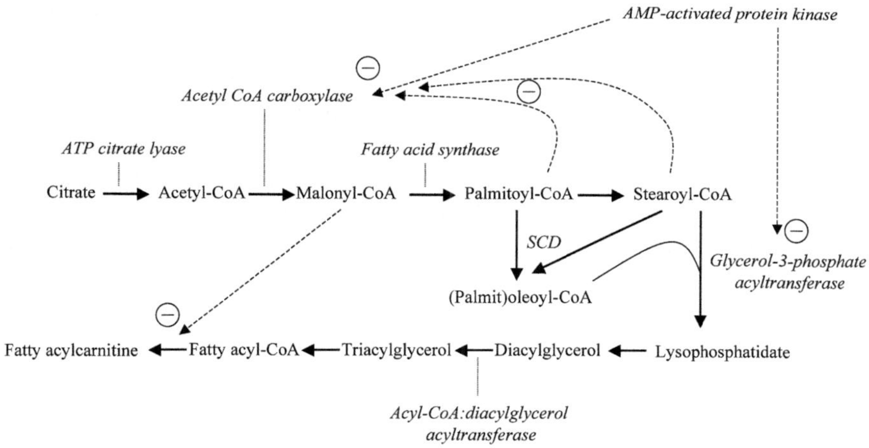

Figure 2. Target enzymes in lipid synthesis. Acetyl-CoA is produced in the mitochondrion and condenses with oxaloacetate to form citrate. Citrate is transported out of the mitochondrion by the tricarboxylate carrier, so that fatty acid and triglyceride synthesis can take place in the cytosol. Cytosolic citrate is the starting point of the scheme. The scheme also shows that malonyl-CoA inhibits the formation of fatty acylcarnitine, thereby preventing the entry of fatty acids into the mitochondrion for oxidation. Potential anti-obesity drugs must inhibit each of the enzymes, except for AMP-activated protein kinase, which inhibits the activity of other enzymes. Abbreviation: SCD – stearoyl-CoA desaturase.

Fatty acid synthesis

Acetyl CoA carboxylase (ACC) produces malonyl-CoA. Malonyl-CoA is used to initiate fatty acid chain building by fatty acid synthase. It also inhibits carnitine palmitoyl transferase 1, which plays an essential role in the transfer of fatty acids into the mitochondrion for oxidation. It appears that

ACC-1, which is found mainly in lipogenic tissues, produces malonyl-CoA that is used for fatty acid synthesis, whereas ACC-2, which is found mainly in energy using tissues, produces malonyl-CoA that regulates fatty acid oxidation. ACC-2 knockout mice are lean, despite being hyperphagic. Selective inhibition of ACC-2 would appear to be the safer therapeutic option, but such compounds have not been described, whereas non-selective inhibitors have been shown to stimulate fatty acid oxidation in muscle cell lines and *in vivo* [50].

The fatty acid synthase inhibitors cerulenin and C75 were originally claimed to inhibit feeding by acting within the brain. The mechanism seemed to involve elevation of hypothalamic malonyl-CoA. (Contrast the benefit of lowering malonyl-CoA in the periphery by inhibiting ACC-2.) Subsequent studies have shown that C75 also promotes fat oxidation, perhaps in part due to a peripheral action, and disconnected the anorectic effect from hypothalamic fatty acid synthase inhibition [51, 52]. Whatever, the mechanism, inhibition of fatty acid synthase seems more likely to produce toxicity than approaches that involve selective inhibition of one isozyme of an enzyme that regulates the concentration of a regulatory molecule.

Stearoyl-CoA desaturase (SCD) is a microsomal enzyme that catalyses the synthesis of monounsaturated fatty acids from saturated fatty acyl-CoAs, especially palmitoyl- and stearoyl-CoA. There are at least four isoforms of SCD in mice and two in humans. Mice that carry a mutant SCD-1 isoform are resistant to obesity and antisense oligonucleotide inhibitors of SCD-1, apparently acting in the liver, prevented diet-induced obesity in mice [53]. One suggestion is that saturated but not monounsaturated fatty acyl-CoAs potently inhibit ACC.

Triacylglycerol synthesis

Glycerol-3-phosphate acyltransferase (GPAT) begins the process of triacylglycerol synthesis by esterifying glycerol-3-phosphate with fatty acyl-CoA to form lysophosphatidate. There are two forms of GPAT – one located in the outer mitochondrial membrane (mtGPAT) and the other in the endoplasmic reticulum (erGPAT). mtGPAT seems to divert fatty acyl-CoA away from oxidation in the mitochondrion and send it in the form of lysophosphatidate to the endoplasmic reticulum to be used for triacylglycerol synthesis. Mice that lack mtGPAT are leaner than wild type mice and hepatic metabolism favours fatty acid oxidation over triacylglycerol synthesis [54]. By contrast, overexpression of GPAT in mouse liver increased triacylglycerol synthesis, reduced fatty acid oxidation and caused marked hepatic steatosis [55].

The last reaction of triacylglycerol synthesis is catalysed by acyl-CoA:diacylglycerol acyltransferase (DGAT). Two isoforms of DGAT are known. DGAT-1 knockout mice are lean and resistant to obesity, despite hyperphagia. They have increased energy expenditure in part due to increased physical activity, but also due to increased fatty acid oxidation in brown adipose and other tissues. This might explain why the concentration of diacylglycerol, the substrate of the enzyme, paradoxically tended to be low in white adipose tissue, skeletal muscle and especially liver of DGAT1 knockout mice. Presumably the absence of DGAT1 in some other tissue promotes diacylglycerol utilisation or inhibits its formation. Since the knockout mice have increased locomotor activity, this other tissue could be the brain [56]. DGAT-2 mice do not survive long after birth, but studies using antisense suggest that inhibitors might reduce hepatic steatosis and hyperlipidaemia.

At the time of writing, full publications have not appeared that show effects of SCD, GPAT or DGAT inhibitors on energy balance. Demonstration of anti-obesity activity with tool compounds would be a major step forward, since the phenotypes of genetically modified mice can be misleading.

Hypothalamic energy metabolism

Initial interest in many of the new intracellular anti-obesity targets has been driven by the phenotypes of mice in which a gene has been knocked out in the whole animal. It might seem that if the gene regulates lipid or carbohydrate metabolism then drugs targeted towards the gene product must act peripherally. The hypothalamus responds to fuels of various kinds, however – lipids, glucose and amino acids – as well as to insulin [4].

In addition to the targets so far described, consider protein tyrosine phosphatase-1B (PTP1B). This enzyme deactivates the insulin receptor and possibly insulin receptor substrates. Inhibitors might be expected to be useful in the treatment of type 2 diabetes, but the PTP1B mouse is also resistant to obesity [57]. Since insulin promotes anabolism when it acts in the periphery, but catabolism when it acts centrally, it seems likely that the lean phenotype of the PTP1B mouse is due to the absence of the enzyme in the brain. Consistent with this view, the neuronal insulin receptor knockout mouse is obese. Another consideration is that PTP1B also deactivates the leptin receptor and Janus-activated kinase-1, which plays a role in leptin signalling. These are also central targets.

The implication of all this is that not only might some of the targets require a centrally acting drug, but others might require drugs that do not penetrate the brain because central and peripheral actions oppose each other. One wants to lower malonyl-CoA in the periphery to stimulate fatty acid oxidation, but increase it centrally to inhibit food intake. Similarly if AMPK activators have potential, they must not reach the hypothalamus because here this would promote food intake. There is logic in these opposing peripheral and central actions of malonyl-CoA and AMPK on energy balance. They both play a role in sensing fuel availability. In times of energy need not only must oxidative metabolism be increased, but food intake must increase to replace the oxidised fuel. On the other hand leptin, which has the role of regulating body weight, activates AMPK in the periphery in order to burn off fuel, whilst inhibiting AMPK in the hypothalamus to prevent the fuel being replaced [58].

Conclusion

Peripherally acting drugs that stimulate energy expenditure or alter substrate utilisation have many attractions, ranging from pragmatism (it is easier to make drugs that do not have to penetrate the CNS), through aetiology (obese subjects oxidise fat poorly), to therapeutic benefit (improved insulin sensitivity). There are many potential targets for such drugs. Some are hormone receptors or alter the release, metabolism or action of hormones. Others are transcription factors or enzymes that regulate mitochondrial biogenesis, or enzymes that regulate lipid metabolism. Surprisingly, some enzymes that regulate fatty acid and triglyceride synthesis seem to alter energy balance primarily by increasing energy expenditure rather than reducing energy intake.

Some of these targets are suggested by mechanisms that are known to influence energy expenditure and may be supported by effects of compounds in rodents and humans. Others are more speculative and supported only by the phenotypes of genetically modified mice. Indeed, the true location of some of the targets may be in the hypothalamus, since hypothalamic metabolism influences both energy intake and expenditure. Moreover, there are examples where centrally and peripherally acting drugs would be expected to have opposing actions on energy balance.

There is no shortage of targets. The difficulty is in deciding which, if any, is likely to deliver a drug.

References

1 Arch JR (2002) Lessons in obesity from transgenic animals. *J Endocrinol Invest* 25: 867–875
2 Clapham JC, Arch JRS, Tadayyon M (2001) Anti-obesity drugs: A critical review of current therapies and future opportunities. *Pharmacol Ther* 89: 81–121
3 Luque CA, Rey JA (2002) The discovery and status of sibutramine as an anti-obesity drug. *Eur J Pharmacol* 440: 119–128
4 Lam TK, Schwartz GJ, Rossetti L (2005) Hypothalamic sensing of fatty acids. *Nat Neurosci* 8: 579–584
5 Ravussin E, Lillioja S, Knowler WC, Christin L, Freymond D, Abbott WG, Boyce V, Howard BV, Bogardus C (1988) Reduced rate of energy expenditure as a risk factor for body-weight gain. *N Engl J Med* 318: 467–472
6 Levine JA, Eberhardt NL, Jensen MD (1999) Role of nonexercise activity thermogenesis in resistance to fat gain in humans. *Science* 283: 212–214
7 Buscemi S, Verga S, Caimi G, Cerasola G (2005) Low relative resting metabolic rate and body weight gain in adult Caucasian Italians. *Int J Obes Relat Metab Disord* 29: 287–291
8 Astrup A, Gotzsche PC, van de Werken K, Ranneries C, Toubro S, Raben A, Buemann B (1999) Meta-analysis of resting metabolic rate in formerly obese subjects. *Am J Clin Nutr* 69: 1117–1122

9 Blaak EE, Wolffenbuttel BH, Saris WH, Pelsers MM, Wagenmakers AJ (2001) Weight reduction and the impaired plasma-derived free fatty acid oxidation in type 2 diabetic subjects. *J Clin Endocrinol Metab* 86: 1638–1644
10 Kim JY, Hickner RC, Cortright RL, Dohm GL, Houmard JA (2000) Lipid oxidation is reduced in obese human skeletal muscle. *Am J Physiol Endocrinol Metab* 279: E1039–1044
11 Tanner CJ, Barakat HA, Dohm GL, Pories WJ, MacDonald KG, Cunningham PR, Swanson MS, Houmard JA (2002) Muscle fiber type is associated with obesity and weight loss. *Am J Physiol Endocrinol Metab* 282: E1191–1196
12 Arch JR (2002) β_3-Adrenoceptor agonists: potential, pitfalls and progress. *Eur J Pharmacol* 440: 99–107
13 Durrant ML, Garrow JS, Royston P, Stalley SF, Sunkin S, Warwick PM (1980) Factors influencing the composition of the weight lost by obese patients on a reducing diet. *Brit J Nutrition* 44: 275–285
14 Darimont C, Turini M, Epitaux M, Zbinden I, Richelle M, Montell E, Ferrer-Martinez A, Mace K (2004) beta3-adrenoceptor agonist prevents alterations of muscle diacylglycerol and adipose tissue phospholipids induced by a cafeteria diet. *Nutr Metab (Lond)* 1: 4
15 Padwal R, Li SK, Lau DC (2003) Long-term pharmacotherapy for overweight and obesity: a systematic review and meta-analysis of randomized controlled trials. *Int J Obes Relat Metab Disord* 27: 1437–1446
16 Schiffelers SLH, Blaak EE, Saris WHM, van Baak MA (2000) *In vivo* β_3-adrenergic stimulation of human thermogenesis and lipid use. *Clin Pharmacol Therapeutics* 67: 558–566
17 Blundell JE, Stubbs RJ, Hughes DA, Whybrow S, King NA (2003) Cross talk between physical activity and appetite control: does physical activity stimulate appetite? *Proc Nutr Soc* 62: 651–661
18 Rolfe DF, Brown GC (1997) Cellular energy utilization and molecular origin of standard metabolic rate in mammals. *Physiol Rev* 77: 731–758
19 Bray GA (1969) Calorigenic effect of human growth hormone in obesity. *J Clin Endocrinol Metab* 29: 119–122
20 Silva JE (2003) The thermogenic effect of thyroid hormone and its clinical implications. *Ann Intern Med* 139: 205–213
21 Brillon DJ, Zheng B, Campbell RG, Matthews DE (1995) Effect of cortisol on energy expenditure and amino acid metabolism in humans. *Am J Physiol* 268: E501–513
22 Silvestri E, Schiavo L, Lombardi A, Goglia F (2005) Thyroid hormones as molecular determinants of thermogenesis. *Acta Physiol Scand* 184: 265–283
23 Grover GJ, Mellstrom K, Ye L, Malm J, Li YL, Bladh LG, Sleph PG, Smith MA, George R, Vennstrom B et al. (2003) Selective thyroid hormone receptor-β activation: a strategy for reduction of weight, cholesterol, and lipoprotein (a) with reduced cardiovascular liability. *Proc Natl Acad Sci USA* 100: 10067–10072
24 Larsen TM, Toubro S, van Baak MA, Gottesdiener KM, Larson P, Saris WH, Astrup A (2002) Effect of a 28-d treatment with L-796568, a novel β_3-adrenergic receptor agonist, on energy expenditure and body composition in obese men. *Am J Clin Nutr* 76: 780–788
25 Greenway FL, De Jonge L, Blanchard D, Frisard M, Smith SR (2004) Effect of a dietary herbal supplement containing caffeine and ephedra on weight, metabolic rate, and body composition. *Obes Res* 12: 1152–1157
26 Russell ST, Hirai K, Tisdale MJ (2002) Role of β_3-adrenergic receptors in the action of a tumour lipid mobilizing factor. *Br J Cancer* 86: 424–428
27 Herd C, Wittert G, Caterson I, Profietto J, Strauss B, Prins J, Stocks A, Vos A, Belyea C (2005) The effect of AOD9604 on weight loss in obese adults: results of a randomized, double-blind, placebo-controlled, multicenter study. *Obesity Res* 13 suppl: A27
28 Proietto J, Thorburn AW (2003) The therapeutic potential of leptin. *Expert Opin Investig Drugs* 12: 373–378
29 Maneuf Y, Higginbottom M, Pritchard M, Lione L, Ashford MLJ, Richardson PJ (2004) Small molecule leptin mimetics overcome leptin resistance in obese rats. *Fundam Clin Pharmacol* 18 Suppl 1: 83
30 Bluher S, Ziotopoulou M, Bullen JW Jr, Moschos SJ, Ungsunan L, Kokkotou E, Maratos-Flier E, Mantzoros CS (2004) Responsiveness to peripherally administered melanocortins in lean and obese mice. *Diabetes* 53: 82–90
31 Kadowaki T, Yamauchi T (2005) Adiponectin and adiponectin receptors. *Endocr Rev* 26: 439–451

32 Morton NM, Paterson JM, Masuzaki H, Holmes MC, Staels B, Fievet C, Walker BR, Flier JS, Mullins JJ, Seckl JR (2004) Novel adipose tissue-mediated resistance to diet-induced visceral obesity in 11 beta-hydroxysteroid dehydrogenase type 1-deficient mice. *Diabetes* 53: 931–938
33 Arch JRS, Wang SJY, Birtles S, Smith DM, Turnbull A (2005) Effects of an inhibitor of 11beta-hydroxysteroid dehydrogenase type 1 inhibitor on energy balance and glucose homeostasis in diet-induced obesity. *Diabetologia* 48: A238
34 Clapham JC (2004) Treating obesity: pharmacology of energy expenditure. *Curr Drug Targets* 5: 309–323
35 Clapham JC, Arch JRS, Chapman H, Haynes AC, Lister CA, Moore GBT, Piercy V, Smith SA, Beeley LJ, Godden RJ et al. (2000) Mice overexpressing human uncoupling protein-3 in skeletal muscle are hyperphagic and lean. *Nature* 406: 415–418
36 Nedergaard J, Cannon B (2003) The 'novel' 'uncoupling' proteins UCP2 and UCP3: what do they really do? Pros and cons for suggested functions. *Exp Physiol* 88: 65–84
37 Ruderman N, Prentki M (2004) AMP kinase and malonyl-CoA: targets for therapy of the metabolic syndrome. *Nat Rev Drug Discov* 3: 340–351
38 Buhl ES, Jessen N, Pold R, Ledet T, Flyvbjerg A, Pedersen SB, Pedersen O, Schmitz O, Lund S (2002) Long-term AICAR administration reduces metabolic disturbances and lowers blood pressure in rats displaying features of the insulin resistance syndrome. *Diabetes* 51: 2199–2206
39 Camacho RC, Pencek RR, Lacy DB, James FD, Donahue EP, Wasserman DH (2005) Portal venous 5-aminoimidazole-4-carboxamide-1-beta-D-ribofuranoside infusion overcomes hyperinsulinemic suppression of endogenous glucose output. *Diabetes* 54: 373–382
40 Puigserver P, Spiegelman BM (2003) Peroxisome proliferator-activated receptor-gamma coactivator 1 alpha (PGC-1 alpha): transcriptional coactivator and metabolic regulator. *Endocr Rev* 24: 78–90
41 Kamei Y, Ohizumi H, Fujitani Y, Nemoto T, Tanaka T, Takahashi N, Kawada T, Miyoshi M, Ezaki O, Kakizuka A (2003) PPARgamma coactivator 1beta/ERR ligand 1 is an ERR protein ligand, whose expression induces a high-energy expenditure and antagonizes obesity. *Proc Natl Acad Sci USA* 100: 12378–12383
42 Willy PJ, Murray IR, Qian J, Busch BB, Stevens WC Jr, Martin R, Mohan R, Zhou S, Ordentlich P, Wei P et al. (2004) Regulation of PPARgamma coactivator 1alpha (PGC-1alpha) signaling by an estrogen-related receptor alpha (ERRalpha) ligand. *Proc Natl Acad Sci USA* 101: 8912–8917
43 Luo J, Sladek R, Carrier J, Bader JA, Richard D, Giguere V (2003) Reduced fat mass in mice lacking orphan nuclear receptor estrogen-related receptor alpha. *Mol Cell Biol* 23: 7947–7956
44 Berger JP, Akiyama TE, Meinke PT (2005) PPARs: therapeutic targets for metabolic disease. *Trends Pharmacol Sci* 26: 244–251
45 Tenenbaum A, Motro M, Fisman EZ (2005) Dual and pan-peroxisome proliferator-activated receptors (PPAR) co-agonism: the bezafibrate lessons. *Cardiovasc Diabetol* 4: 14
46 Tiraby C, Langin D (2003) Conversion from white to brown adipocytes: a strategy for the control of fat mass? *Trends Endocrinol Metab* 14: 439–441
47 Obici S, Rossetti L (2003) Minireview: nutrient sensing and the regulation of insulin action and energy balance. *Endocrinology* 144: 5172–5178
48 Westerterp-Plantenga MS, Kovacs EM (2002) The effect of (–)-hydroxycitrate on energy intake and satiety in overweight humans. *Int J Obes Relat Metab Disord* 26: 870–872
49 Ohia SE, Opere CA, LeDay AM, Bagchi M, Bagchi D, Stohs SJ (2002) Safety and mechanism of appetite suppression by a novel hydroxycitric acid extract (HCA-SX). *Mol Cell Biochem* 238: 89–103
50 Harwood HJ Jr (2004) Acetyl-CoA carboxylase inhibition for the treatment of metabolic syndrome. *Curr Opin Investig Drugs* 5: 283–289
51 Tu Y, Thupari JN, Kim EK, Pinn ML, Moran TH, Ronnett GV, Kuhajda FP (2005) C75 alters central and peripheral gene expression to reduce food intake and increase energy expenditure. *Endocrinology* 146: 486–493
52 Rohrbach KW, Han S, Gan J, O'Tanyi EJ, Zhang H, Chi CL, Taub R, Largent BL, Cheng D (2005) Disconnection between the early onset anorectic effects by C75 and hypothalamic fatty acid synthase inhibition in rodents. *Eur J Pharmacol* 511: 31–41
53 Jiang G, Li Z, Liu F, Ellsworth K, Dallas-Yang Q, Wu M, Ronan J, Esau C, Murphy C, Szalkowski D et al. (2005) Prevention of obesity in mice by antisense oligonucleotide inhibitors of stearoyl-CoA desaturase-1. *J Clin Invest* 115: 1030–1038
54 Thuresson ER (2004) Inhibition of glycerol-3-phosphate acyltransferase as a potential treatment

for insulin resistance and type 2 diabetes. *Curr Opin Investig Drugs* 5: 411–418
55 Lindén D, William-Olsson L, Ahnmark A, Ekroos K, Hallberg C, Peilot-Sjögren H, Becker B, Svensson L, Clapham JC, Oscarsson J et al. (2006) Liver directed overexpression of mitochondrial glycerol-3-phosphate acyltransferase results in hepatic steatosis, increased triglyceride secretion and reduced fatty acid oxidation. *FASEB Journal* 20: 434–443
56 Chen HC, Farese RV Jr (2005) Inhibition of triglyceride synthesis as a treatment strategy for obesity: lessons from DGAT1-deficient mice. *Arterioscler Thromb Vasc Biol* 25: 482–486
57 Taylor SD, Hill B (2004) Recent advances in protein tyrosine phosphatase 1B inhibitors. *Expert Opin Investig Drugs* 13: 199–214
58 Kim MS, Lee KU (2005) Role of hypothalamic 5'-AMP-activated protein kinase in the regulation of food intake and energy homeostasis. *J Mol Med* 83: 514–520

Index

The MDT-Series
Milestones in Drug Therapy

The discovery of drugs is still an unpredictable process. Breakthroughs are often the result of a combination of factors, including serendipidity, rational strategies and a few individuals with novel ideas. *Milestones in Drug Therapy* highlights new therapeutic developments that have provided significant steps forward in the fight against disease. Each book deals with an individual drug or drug class that has altered the approach to therapy. Emphasis is placed on the scientific background to the discoveries and the development of the therapy, with an overview of the current state of knowledge provided by experts in the field, revealing also the personal stories behind these milestone developments. The series is aimed at a broad readership, covering biotechnology, biochemistry, pharmacology and clinical therapy.

Forthcoming titles

Treatment of Psoriasis, J.M. Weinberg (Editor), 2008
Echinocandin Antifungals, K. Bartizal, E. Hickey (Editors), 2008
Bipolar Depression: Molecular Neurobiology, Clinical Diagnosis and Pharmacotherapy, C.A. Zarate, H.K. Manji (Editors), 2008

Published volumes

Entry Inhibitors in HIV Therapy, J.D. Reeves, C.A. Derdeyn (Editors), 2007
Drugs affecting Growth of Tumours, H.M. Pinedo, C. Smorenburg (Editors), 2006
TNF-alpha Inhibitors, J.M. Weinberg, R. Buchholz (Editors), 2006
Aromatase Inhibitors, B.J.A. Furr (Editor), 2006
Cannabinoids as Therapeutics, R. Mechoulam (Editor), 2005
St. John`s Wort and its Active Principles in Anxiety and Depression, W.E. Müller (Editor), 2005
Drugs for Relapse Prevention of Alcoholism, R. Spanagel, K. Mann (Editors), 2005
COX-2 Inhibitors, M. Pairet, J. Van Ryn (Editors), 2004
Calcium Channel Blockers, T. Godfraind (Author), 2004
Sildenafil, U. Dunzendorfer (Editor), 2004
Hepatitis Prevention and Treatment, J. Colacino, B.A. Heinz (Editors), 2004
Combination Therapy of AIDS, E. De Clercq, A.M. Vandamme (Editors), 2004
Cognitive Enhancing Drugs, J. Buccafusco (Editor), 2004
Fluoroquinolone Antibiotics, A.R. Ronald, D. Low (Editors), 2003

Erythropoietins and Erythropoiesis, G. Molineux, M. Foote, S. Elliott (Editors), 2003
Macrolide Antibiotics, W. Schönfeld, H. Kirst (Editors), 2002
HMG CoA Reduktase Inhibitors, G. Schmitz, M. Torzewski (Editors), 2002
Antidepressants, B.E. Leonard (Editor), 2001
Recombinant Protein Drugs, P. Buckel (Editor), 2001
Glucocorticoids, N. Goulding, R.J. Flower (Editors), 2001
Modern Immunosuppressives, H.-J. Schuurman (Editor), 2001
ACE Inhibitors, P. D'Orleans-Juste, G. Plante (Editors), 2001
Atypical Antipsychotics, A.R. Cools, B.A. Ellenbroek (Editors), 2000
Methotrexate, B.N. Cronstein, J.R. Bertino (Editors), 2000
Anxiolytics, M. Briley, D. Nutt (Editors), 2000
Proton Pump Inhibitors, L. Olbe (Editor), 1999
Valproate, W. Löscher (Editor), 1999

GPSR Compliance

The European Union's (EU) General Product Safety Regulation (GPSR) is a set of rules that requires consumer products to be safe and our obligations to ensure this.

If you have any concerns about our products, you can contact us on ProductSafety@springernature.com

In case Publisher is established outside the EU, the EU authorized representative is:

Springer Nature Customer Service Center GmbH
Europaplatz 3
69115 Heidelberg, Germany

Batch number: 10578049

Printed by Printforce, the Netherlands